SYSTEMATIC BOTANY MONOGRAPHS

VOLUME 59

Revision of Gasteranthus (Gesneriaceae)

Laurence E. Skog
Lars P. Kvist

THE AMERICAN SOCIETY OF PLANT TAXONOMISTS
17 July 2000

SYSTEMATIC BOTANY MONOGRAPHS
ISSN 0737-8211

ISBN 0-912861-59-2

Printed in the United States of America

REVISION OF GASTERANTHUS (GESNERIACEAE)

Laurence E. Skog
Department of Botany, MRC-166
U.S. National Museum of Natural History
Smithsonian Institution
Washington, D.C. 20560-0166, U.S.A.

Lars P. Kvist
Unit of Forestry
Royal Veterinary and Agricultural University
DK-1871 Frederiksberg C (Copenhagen), Denmark

ABSTRACT. *Gasteranthus* (Gesneriaceae-Beslerieae) comprises 35 species, including three subspecies. Its range extends from Guatemala and nearby Mexico, Costa Rica, and Panama through western South America south to Bolivia. The center of diversity is in western Ecuador where 25 species occur. Species of *Gasteranthus* are terrestrial herbs or short subshrubs, and display a diverse array of corolla variation. Corollas may be pale, whitish and funnelform with a large limb and patterns of darker dots and lines inside the throat, or may be relatively uniformly bright yellow, orange, or red, and urceolate with an often strongly inflated ventral pouch and a small throat and limb; intermediate corolla shapes have a ventricose, subventricose, or funnelform tube with a small limb. Species of *Gasteranthus* grow in forests throughout the range, mainly in shady and permanently humid places, usually in ravines with streams or near waterfalls, but particularly in cloud forests in the permanently wet understory vegetation. The high diversity of *Gasteranthus* in western Ecuador apparently reflects the speciation of the genus in separate low- and high-elevation cloud forests. Ten species of *Gasteranthus* are found predominantly in the low-elevation forests between 500 and 800 m elevation, and another 10 species are restricted to the higher forests above 1,800 m elevation. The remaining species appear to be less habitat-specific and consequently more widespread. Because of their limited distribution in cloud forests, the species of *Gasteranthus* are vulnerable to extinction. Particularly, the low-elevation cloud forests have been nearly completely destroyed; at least 5 endemic species already may be extinct in western Ecuador and an additional 7 throughout the entire range of *Gasteranthus*. Due to loss of habitat, the survival of more than 10 other species of *Gasteranthus* may be questionable. Eleven species are here newly described: *G. adenocalyx, G. bilsaensis, G. epedunculatus, G. extinctus, G. glaber, G. mutabilis, G. orientandinus, G. osaensis, G. tenellus, G. ternatus,* and *G. villosus.* Two new combinations are proposed: *G. calcaratus* subsp. *calceolus* and *G. calcaratus* subsp *oncogastrus*.

INTRODUCTION

Three stimuli provided us with the impetus to revise *Gasteranthus*: first, the attractiveness of the flowers. Species of *Gasteranthus* are among the most remarkable New World Gesneriaceae, with their large brightly colored and peculiarly shaped corollas, which suggest highly specific co-evolved flower-pollinator interactions. Second, a revision of the genus was long overdue and necessary before a treatment could be prepared for the Flora of Ecuador project. The third stimulus for the present work is the existence of 16 species of *Gasteranthus* that are endemic to western Ecuador (Appendix). Indeed, six species have been reported to be endemic to a single, 600 m high front-ridge of the western Andean slopes known as Centinela Ridge (Gentry & Dodson 1987; Gentry 1989, 1992; Dodson & Gentry 1991). It is clear to us that such narrowly endemic species are

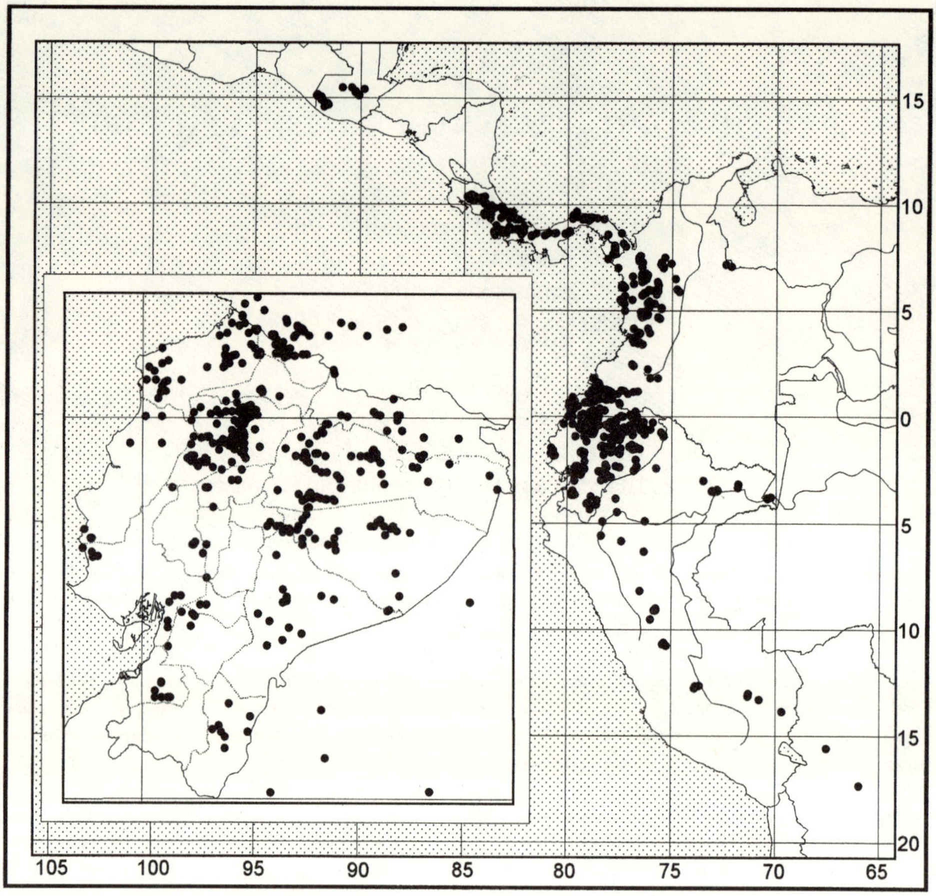

FIG. 1. Distribution of *Gasteranthus* in Central and South America. Inset: distribution of *Gasteranthus* in Ecuador.

highly endangered if not already extinct (Kvist et al., unpubl.), because much of the forest of western Ecuador has been converted to farmland (Dodson & Gentry 1991; Parker & Carr 1992).

Gasteranthus, a medium-sized genus in the neotropical subfamily Gesnerioideae of the Gesneriaceae, occurs from Mexico south to Bolivia (Fig. 1). The genus is related to *Besleria* L. and *Cremosperma* Benth. in the Beslerieae, a tribe characterized by the lack of bracts in the inflorescences. *Gasteranthus* can be distinguished from the other two large genera in the tribe by several traits summarized in Table 1, but the presence of the stomatal clusters on the abaxial leaf surface is an obvious generic character. We recognize thirty-five species, including three subspecies, of *Gasteranthus*. Plants are frequently found in wet areas in wet montane forests and often near streams. Remarkably, the plants are adapted to dense shade of the forest understory where few other plants thrive. The corolla is large and/or colorful in most species in hues of yellow, orange, or red, and the white corollas of some species often have lines or spots. Unfortunately, about half of the species are presumed to be extinct or endangered.

Morton (1939) revised *Gasteranthus* as a section of *Besleria*, but his work is outdated, and does not allow identification of many of the species currently referred to *Gasteranthus*. Biological studies and conservation efforts depend on proper identification and circumscription of the variation found throughout the ranges of the individual species; the present study intends to provide these much-needed data, and to encourage urgently needed research and conservation efforts.

The second author first visited western Ecuador and collected Gesneriaceae in 1982 and 1983. Both authors visited western Ecuador and collected Gesneriaceae in October 1995 with Paulina Mendoza-T. and others. In addition, the second author also visited Colombia and collected in the department of Chocó. During these and other trips, the authors have studied 17 of the 35 species in the field. During the 1995 trip the authors also studied the collections in the herbaria in Bogotá (COL), Medellín (HUA), and in Quito (QCA, QCNE), to supplement the many hundreds of specimens on loan for study from several herbaria. Several of the species of *Gasteranthus* have been in cultivation in the greenhouses of the Smithsonian Institution for study and illustration for this treatment.

TAXONOMIC HISTORY

Gasteranthus (Gesneriaceae) was described by Bentham (1846) based on a Hartweg collection from Ecuador that he named *Gasteranthus quitensis*; however, the earliest described species that is now included in *Gasteranthus* is *G. calcaratus* (Kunth) Wiehler, originally described by Kunth in 1818 as *Besleria calcarata*. Hanstein (1865) and then Bentham himself (1876) recognized *Gasteranthus* as a subgenus and a section of *Besleria* L., respectively. Morton (1939) revised *Besleria* and treated *Gasteranthus* as one of four sections of *Besleria*. In the section *Gasteranthus* Morton included species with strongly urceolate corollas, while he assigned the other species now included in the genus *Gasteranthus* to section *Neobesleria*. The latter section was distinguished by a dorsally thickened nectary and a spur (but not a strongly inflated corolla with a ventral pouch as in the otherwise similar section *Gasteranthus*). Species now included in *Gasteranthus* have also been described as species of *Besleria* in the works by the following authors: Fritsch (1922, 1934); Morton (1935, 1944, 1968); J. D. Smith (1891, 1912); and Standley and Williams (1951). In addition, Seemann (1854), Brandegee (1917), and Mansfeld (1936) each described one currently accepted species of *Gasteranthus*, but did so in the genera *Drymonia* Martius, *Episcia* Martius, and *Halphophyllum* Mansf., respectively. Wiehler reestablished *Gasteranthus* as a genus in 1975, and described an additional five new species of *Gasteranthus*. More recently, Skog and Kvist (1994) and Freiberg (1996, 1998) described new species of *Gasteranthus*.

GENERIC AND SPECIFIC DELIMITATION

Wiehler (1983) and Burtt and Wiehler (1995) placed *Gasteranthus* next to the large genus *Besleria* in the tribe Beslerieae, which is defined by the absence of bracts in the inflorescences (Wiehler 1983). *Besleria* and *Cremosperma* are the other two large genera in Beslerieae with ca. 150 and 30 species, respectively. Species of *Besleria* and *Cremosperma* can often be found in the same habitats as those of *Gasteranthus*, but the three genera are relatively easy to distinguish to genus by the characters listed in Table 1, especially when fully developed flowers or fruits are present. The often-conspicuous stomatal clusters on the lower leaf surfaces can easily identify sterile specimens of *Gasteranthus*.

TABLE 1. Comparison of *Gasteranthus*, *Besleria*, and *Cremosperma*.

	Gasteranthus	*Besleria*	*Cremosperma*
Habit	herbs/subshrubs	subshrubs/shrubs	herbs
Stomata	clustered	scattered	scattered
Inflorescence	usually pedunculate	mostly epedunculate	pedunculate
Calyx lobes	nearly free	nearly free	adnate half of length
Calyx spur	usually present	sometimes present	absent
Attachment of filaments	usually only basally	basally	attached to lower half of corolla
Nectary	dorsal to semiannular	annular	dorsal to annular
Fruit	fleshy or semi-fleshy capsule	berry	semi-fleshy capsule
Seeds	broadly elliptic	broadly elliptic	narrowly elliptic

Conspicuous stomatal clusters are apparently not found in other New World Gesneriaceae, but stomatal clusters have been reported from *Reldia* Wiehler, *Napeanthus* Gardner, and *Gesneria* L. Yet, Kvist and Skog (1989) found no detectable stomatal clusters in *Reldia*, which Wiehler also placed in the Beslerieae; it differs from *Gasteranthus*, *Besleria*, and *Cremosperma* chiefly by having alternate leaves. Another genus, *Resia* H. E. Moore, was included in Beslerieae by Wiehler (1983) and by Burtt and Wiehler (1995). A recent study (Skog & de Jesús 1997) has shown that *Resia* does not belong in Beslerieae, if the tribe is strictly defined by the absence of bracts, since a newly described subspecies (*Resia ichthyoides* Leeuwenb. subsp. *bracteata* L. E. Skog & de Jesús) has bracteate inflorescences. *Resia* is probably more closely related to the genus *Napeanthus*, and belongs to the previously monogeneric tribe Napeantheae (Skog & de Jesús 1997). Finally, according to Burtt and Wiehler (1995) the tribe Beslerieae also includes three unusual and little-known genera, *Anetanthus* Hiern ex Benth., *Cubitanthus* Barringer, and *Tylopsacas* Leeuwenb., with two, one, and one species, respectively. It remains to be yet proven if these small genera truly are related to the larger tribal core-genera, *Gasteranthus*, *Besleria*, and *Cremosperma*.

When reestablishing *Gasteranthus*, Wiehler (1975) distinguished *Gasteranthus* from *Besleria* by characters of the leaf stomatal clusters, the nectary, and the fruit. We use these and additional characters in combination to define *Gasteranthus* and to distinguish it from *Besleria* and *Cremosperma* in the tribe Beslerieae (see Table 1), and from all other New World Gesneriaceae: 1) stomata in clusters on the lower leaf surfaces, which usually but not always are conspicuous to the naked eye; 2) inflorescences without floral bracts; 3) nectaries reduced to a dorsal bilobed or somewhat irregular gland, or at least the nectary thicker and higher dorsally than ventrally rather than perfectly annular (as in most species of *Besleria*); 4) the fruit a compressed fleshy or semi-fleshy capsule (less fleshy than in *Alloplectus* Martius) that exposes the seeds by splitting with two reflexing valves, and which may split secondarily into four valves; and 5) seeds less than 0.5 mm long and without apparent funiculi. In addition, almost all species of *Gasteranthus* have pedunculate inflorescences and corollas with a spur. Only *G. epedunculatus* is both epedunculate and ecalcarate.

Originally Bentham (1846) established *Gasteranthus* based mainly on the distinctive urceolate corollas of *G. quitensis*, which have strongly inflated, mostly protruding pouches (Fig. 4D), in contrast to the corollas found in *Besleria*, which are mostly tubular and narrowly funnelform. Only approximately a third of the species currently included in

Gasteranthus have this distinctive urceolate corolla type. Morton (1939) recognized another subgroup in *Besleria* that equally represents about a third of the present *Gasteranthus* species and named it section *Neobesleria*. The species that Morton included in his sect. *Neobesleria* tend to have large funnelform to campanulate whitish corollas with a large spur, which differ from the smaller, bright yellow, orange, or red corollas of *Besleria*. Wiehler (1975, 1978) proposed that the groups of species with urceolate and funnelform corollas conform to the ornithophilous and the euglossophilous pollination-syndromes, respectively. Freiberg (1998) also recognized these two subgroups in *Gasteranthus*. According to him, they could be considered as sections in the genus, although "the still rising number of species and their local distributions makes it difficult to subdivide *Gasteranthus*." In addition, Freiberg (1998) mentions that one of his new species, *G. otongensis*, has characters somewhat intermediate between his two potential sections; it appears to have a small pouch below the corolla while its throat is not constricted. However, *G. otongensis* is far from the only species that does not fit well into one of the two potential subgeneric taxa. There exists as well a third species group with relatively small, brightly colored, ventricose corollas (see Table 2, Group B). These species are less distinct from *Besleria* than those included in the two subgroups with more conspicuous flowers.

All 35 species of *Gasteranthus* recognized in this work are listed in Table 2 sorted into three groups. The first group (A) includes species having mostly funnelform (euglossophilous) corollas without a pouch, corresponding approximately to *Besleria* sect. *Neobesleria* (Morton 1939). The second group (B) includes species having mostly ventricose corollas with more or less well-developed ventral pouches and terminal corolla throats. The third group (C) includes species having urceolate (ornithophilous) corollas with a large protruding pouch and a dorsal corolla throat, and includes the type species *Gasteranthus quitensis*. The delimitation of the three species groups is difficult; e.g., corolla shapes and colors do not always correlate with the state of the pouch. *Gasteranthus corallinus* (Group B) and three other species have urceolate corollas without protruding pouches, *G. acropodus* and *G. villosus* (both Group A) have ventricose corollas without pouches, and *G. columbianus* (Group A) has funnelform bright (red) corollas. Table 2 ignores the intra-specific variation of some species, implying that the species group B with somewhat *Besleria*-like flowers spans the variation spectrum between the two more distinct species groups with the larger funnelform (A) and the urceolate corollas (C). Because of the considerable variation in corolla form we have concluded that *Gasteranthus* should not be divided into two (or even three) sections or at least a classification should not be based solely on corolla characters. In addition, we believe that *Gasteranthus* in its present form seems to be a monophyletic group developed from an ancestor within or near *Besleria*.

Some species of *Gasteranthus* are similar in habit to species of *Besleria*, e.g., *G. carinatus*, *G. timidus*, *G. columbianus*, and *G. imbricans*, which are relatively shrubby and superficially similar to the widespread *Besleria solanoides* Kunth and species related to it. *Gasteranthus* may have originated within its present diversity center along the western Andean slopes, although *G. imbricans* actually occurs in Panama and Costa Rica. None of the three species groups noted in Table 2 has a particularly limited distribution. All three groups include widespread species, particularly *G. wendlandianus* (Group A, corollas without a pouch), *G. corallinus* (Group B, corollas without a protruding pouch), and *G. calcaratus* and *G. pansamalanus* (Group C, corollas with a protruding pouch). None of

TABLE 2. Survey of selected corolla characters that may influence the pollination biology of *Gasteranthus*. Group A: corollas without a pouch; Group B: corollas with a ventral pouch and a terminal throat; Group C: corollas with a ventral protruding pouch giving the throat a dorsal position. Eight corolla characters are surveyed: shape, color (light vs. bright), and relative size of corolla; character of apex (pointed vs. blunt), and direction and relative length of spur; presence or absence of darker markings (+/-) and indumentum (+/-) in the corolla throat. The numbers to the left refer to illustrations of corollas and distribution maps of the individual species.

Corolla-characters:		Corolla			Spur			Throat		Illu.	Map
		Shape	Colors	size	Apex	Direction	Length	Mar.	Indu.	No.	No.
A. No pouch present:											
11	delphinioides	Funnelform	Light	Large	Pointed	Down	Long	+	+	3B	24
30	tenellus	"	"	"	"	"	"	+	+	3C	24
4	atratus	"	"	"	"	"	"	+	-	3A	24
2	adenocalyx	"	"	"	"	"	Medium	+	+	2B	15
12	dressleri	"	"	"	Blunt	"	Long	+	+	3D	28
23	orientandinus	"	"	"	"	"	"	+	+	2D	15
16	herbaceus	"	"	"	"	"	Medium	+	+	3E	28
20	leopardus	"	"	"	"	"	"	+	-	9B	40
13	epedunculatus	"	"	Medium		Ecalcarate		+	+	2A	15
22	mutabilis	"	"	"	"	Straight	Small	+	+	2C	15
35	wendlandianus	"	"	Small	"	"	"	+	-	9A	40
8	columbianus	"	Bright	"	"	"	Medium	-	+	8B	20
1	acropodus	Ventricose	Light	Medium	"	Down	"	+	+	9C	15
34	villosus	"	"	"	"	"	"	+	+	9D	15
B. Pouch not protruding:											
3	anomalus	Ventricose	Light	Small		Ecalcarate		-	+	10B	19
17	imbaburensis	"	"		Blunt	Straight	Medium	-	?	7B	20
24	osaensis	"	"	"	"	Down	Small	-	+	10A	20
7	carinatus	"	"	"	"	"	"	-	-	8A	19
25	otongensis	"	"	"	"	Straight	Small	-	+	*	23
32	timidus	"	"	"	"	"	"	-	-	8D	20
18	imbricans	"	"	"	"	Up	Medium	-	+	8C	20
29	recurvatus	Urceolate	"	Medium	"	"	Large	-	-	10C	19
15	glaber	"	"	Small	"	Straight	Small	+	-	7C	19
5	bilsaensis	"	"	"	"	"	"	-	-	7D	22
9	corallinus	"	"	"	"	"	"	-	-	10D	22
C. Pouch protruding:											
10	crispus	Urceolate	Light	Medium	Blunt	Straight	Medium	-	+	5A	23
14	extinctus	"	"	"	"	"	"	-	+	5B	23
6	calcaratus	"	"	"	"	"	"	-	-	6	18
21	macrocalyx	"	"	"	"	"	Small	-	+	4B	29
27	perennis	"	"	"	"	"	"	-	+	4C	35
19	lateralis	"	"	"	"	"	"	-	-	4A	29
26	pansamalanus	"	"	"	"	"	"	-	-	5D	34
31	ternatus	"	"	"	"	"	"	-	-	5C	23
33	trifoliatus	"	"	"	"	"	"	-	-	7A	23
28	quitensis	"	"	Small	"	"	"	-	+	4D	35

*Specimen not available for study, see illustration (Fig. 2B, p. 171) in Freiberg, 1998.

the species with small ventricose to tubular corollas similar to those found in *Besleria* is widely distributed.

Table 2 shows that most species of *Gasteranthus* may be set apart by different combinations of easily detectable corolla characteristics of tube, spur, and throat, which all may influence the behavior and/or success of potential pollinators. Species of

Gasteranthus with urceolate corollas apparently vary less in their floral characteristics than species with funnelform and ventricose corollas, especially in their similar (i.e., short, sturdy, and straight) spurs. Several sympatric species have similar urceolate corollas, e.g., *G. crispus* and *G. extinctus*, as well as *G. lateralis*, and *G. pansamalanus* together with *G. ternatus* and *G. trifoliatus* in Pichincha and Imbabura provinces of Ecuador, respectively. In contrast, species with ventricose or funnelform corollas tend to have allopatric distributions, e.g., *G. atratus* and *G. tenellus*, as well as *G. acropodus* and *G. villosus* (all Group A), thus ruling out natural hybridization between them. The fact that the species rarely hybridize suggests that other features may provide them reproductive isolation.

Genera of New World Gesneriaceae that have been recently revised [*Reldia* (Kvist & Skog 1989), *Heppiella* Regel (Kvist 1990), *Kohleria* Regel (Kvist & Skog 1992), *Columnea* L. (Kvist & Skog 1993), and *Pearcea* Regel (Kvist & Skog 1996)] all include one or a few widely distributed species that are often very variable and difficult to circumscribe, along with a larger number of species that are endemic to relatively small areas and have relatively stable combinations of characteristics. This pattern is repeated to some extent in *Gasteranthus*, but the species of *Gasteranthus* have proven generally less difficult to delimit than those of the above-mentioned genera. The most variable species, *G. calcaratus*, is represented by three geographically disjunct subspecies, which combined are widely distributed and common in Ecuador and Colombia and in Peru (Fig. 18). The western Ecuadorian subsp. *oncogastrus*, in particular, has been confused with several other species. *Gasteranthus calcaratus* is the only species in which infraspecific taxa are recognized. The other three relatively widespread species in the genus, *G. wendlandianus*, *G. pansamalanus*, and *G. corallinus* are all morphologically more consistent. Populations of *G. wendlandianus* from Central America and from southern Peru (Fig. 40) have few apparent morphological differences, and the widely disjunct populations of *G. pansamalanus* (Fig. 34) from Central America and South America are also remarkably similar. Populations of *G. corallinus* (Fig. 22) found along the western slopes of the Andes and in the Amazon Basin are similar except that the plants in the Andean population on average apparently grow taller. The relatively consistent species of *Gasteranthus* may reflect that this genus is less prone to hybridization than, for instance, species in the genera *Kohleria* and *Pearcea*, in which hybridization, backcrossing, and introgression apparently are fairly common phenomena (Kvist & Skog 1992, 1996). In *Gasteranthus* there are only indications of extensive hybridization in one locality involving the widespread and usually stable species *G. wendlandianus* and the more local species *G. leopardus*. The latter species has much larger flowers than the former, but nearly the whole continuum of variation can be found in one Colombian locality (see *G. leopardus*, no. 20).

MORPHOLOGY

Habit. Species of *Gasteranthus* are herbs or subshrubs, and the individual, erect shoots are mostly unbranched or sparsely branched. *Gasteranthus ternatus* may be particularly highly branched. The shoots arise from roots or decumbent stems, and thus the plants frequently grow in small clones. Most species attain heights between 50 cm and 2 m, but according to labels of a few vouchers some species may grow up to 4 or 5 m tall. This information may be erroneous, as well as information claiming that plants were growing as epiphytes. The latter may have been growing on moss-covered trunks somewhat above ground-level, but true epiphytes do not occur in *Gasteranthus*. Some species of *Gasteranthus*, particularly *G. pansamalanus* and *G. ternatus*, may be somewhat scandent, but the genus has no truly climbing species. A few species, *G. anomalus*, *G. dressleri*, and some populations of *G. herbaceus*, have very short shoots with densely congested leaves appressed to the ground. In contrast, *G. atratus* and *G. crispus* do have well-developed internodes, but their leaves may develop from decumbent stems and thus also be appressed to the ground. The internodes of some species have a conspicuously irregularly wrinkled bark, particularly on the older internodes, which also tend to be glabrescent. The wrinkled bark is most conspicuous in *G. carinatus* and *G. imbricans*, both of which can have fairly robust plants with glabrous stems.

Indumentum. Gesneriaceae often have a "hairy look," but despite the conspicuous indumentum, most genera have only a few types of trichomes. Species of *Gasteranthus* have predominantly uniseriate, fairly slender, and long (0.4–1 mm long) trichomes of up to 10 cells, but occasionally these relatively large trichomes also have an apical, glandular cell. Glandular trichomes may also be minute (less than 0.2 mm long), reduced to the glandular cell and a stalk of maximally one or two cells; these are usually located inside the corolla. Finally, equally minute trichomes, forming a puberulent indumentum, are common on corollas and calyces. Despite the fact that trichomes show little variation, indumentum-characters can be useful specific characters. *Gasteranthus extinctus* (Fig. 26), *G. mutabilis* (Fig. 31), and *G. villosus* (Fig. 38) thus have a conspicuous pilose to villous indumentum on stems and inflorescences, whereas most species have less conspicuous appressed indumentum. *Gasteranthus carinatus*, *G. glaber* (Fig. 27), and *G. imbricans* are glabrous, except that the last two species have minute glandular trichomes inside the calyx and corolla, and nectaries that are sparsely pubescent or villous, respectively. *Gasteranthus imbaburensis* (Fig. 7B), *G. osaensis* (Fig. 33), and *G. timidus* (Fig. 8D) are characterized by having the corolla exterior bearing a conspicuous pilose or villous indumentum, partly composed of large glandular trichomes.

Leaves. The normal leaf arrangement is opposite and decussate, but *G. ternatus* and *G. trifoliatus* usually have three or four leaves at a node, whereas *G. perennis*, and in rare cases *G. calcaratus*, may also have ternate leaves. The leaves of all species are isophyllous or somewhat anisophyllous, and only *G. calcaratus* subsp. *calceolus* may have strongly anisophyllous leaves. The shape of the blade is typically ovate, elliptic, obovate, or oblanceolate; however, some species have distinctive leaf shapes, notably *G. perennis* with narrowly lanceolate to oblanceolate blades, and some specimens of *G. trifoliatus* and *G. crispus* with obovate to nearly orbicular blades. The blades have from 5 to 17 secondary veins per side, but the variation is usually much smaller within the individual species, and the number of secondary veins may help to separate similar species, e.g., *G. acropodus* and *G. villosus*. The upper leaf surface is flat, i.e., not rugose (and glabrous or glabrescent) in most species, but strongly rugose in *G. atratus* and *G. bilsaensis* (Fig. 17) and somewhat rugose in *G. dressleri* and *G. quitensis*. The lower leaf surfaces of about a third of the species are characterized by having stomata in clusters appearing to the naked eye as lighter colored dots. In the remaining species, the clusters are too inconspicuous to be easily visible.

Inflorescences. The inflorescences are borne in the upper leaf axils, and are pair-flowered cymes similar to those found in many other Gesneriaceae (Weber 1973, 1978, 1982;

Wiehler 1983; Kvist & Skog 1992, 1996), except that bracts are absent. Most species have peduncles considerably longer than the pedicels and may bear at least 4 and often up to 6, 10, or 16 flowers per inflorescence. *Gasteranthus bilsaensis* (Fig. 17) and *G. corallinus* (Fig. 21) may have inflorescences with up to 20 flowers, *G. calcaratus* subsp. *oncogastrus* with up to 25 flowers, and *G. carinatus* 30 or even more flowers. The inflorescences, however, are usually composed of considerably fewer flowers than the apparent maximum number for the particular species. The flowers in the inflorescences may be densely congested or relatively well separated, depending on the length of the pedicels. Some species always have inflorescences with few flowers, but still with the peduncle longer than the pedicels. *Gasteranthus mutabilis* and *G. ternatus* have maximally two (rarely three) flowers, and *G. pansamalanus* and *G. perennis* have pedunculate inflorescences with only a single flower. Only two species, *G. anomalus* and *G. dressleri*, have peduncles shorter than the pedicels, and *G. herbaceus* and *G. epedunculatus* (Fig. 25) have epedunculate inflorescences. The last two species, however, occasionally have two or three pedicellate flowers in the leaf axils.

Calyx. The calyx lobes are connate only basally and are mostly subequal, but may be unequal differing primarily in size rather than in shape. The dorsal calyx lobe is usually forced backward below and surrounding the corolla spur (Figs. 2B, D, 9), and, in contrast to the other four lobes, usually has a conspicuous keel also, reflecting the presence of the corolla spur. The lateral and ventral lobes are also appressed to the corolla but directed upward and forward, respectively. In the two ecalcarate species, *G. anomalus* (Fig. 10B) and *G. epedunculatus* (Figs. 2A, 25), the five calyx lobes are directed forward or erect, respectively. The shape of the calyx lobes is mostly ovate with an acute apex or lanceolate with an acuminate apex; however, the shape can vary from nearly semiorbicular in *G. glaber* (Fig. 27) to narrowly lanceolate in *G. macrocalyx*, *G. otongensis*, and *G. quitensis* (Fig. 4D). The margin of the calyx lobes can also provide useful taxonomic characters, and particularly *G. leopardus* (Figs. 9B, 30) and *G. wendlandianus* (Figs. 9A, 39) are distinctive by having a conspicuously toothed calyx margin.

Corolla. Few genera of Gesneriaceae (and other flowering plant families) have the spectrum of corolla shapes as diverse as found in *Gasteranthus* (Table 2). *Gasteranthus atratus* and *G. crispus* may exemplify the extremes. *Gasteranthus atratus* has large whitish, campanulate-funnelform corollas with a wide limb and throat, and a long-pointed and downward directed spur (Fig. 3A). *Gasteranthus crispus* has medium-sized red to orange, urceolate corollas with an inflated pouch that protrudes beyond the small throat and limb, and the spur is short and straight (Fig. 5A). Ten species of *Gasteranthus* have corollas relatively similar to *G. atratus* (Figs. 2, 3) and ten species are similar to *G. crispus* (Figs. 4, 5, 6); the remaining 15 species have intermediate forms of corollas more or less constituting a continuum between the extremes (Table 2; Figs. 7, 8, 9, 10). Whitish corollas always have darker dots or lines in the throat, whereas the yellow, orange, or red corollas usually are relatively uniformly colored without obvious lines or spots on the throat or limb (Table 2; Figs. 4, 5, 6, 8, 10). An exception is *G. glaber* (Figs. 7C, 27), which has corollas that are red with a black limb, but mostly only the black limb is seen because the tube is hidden by the surrounding calyx. The vast majority of species of *Gasteranthus* have conspicuous spurs that provide useful taxonomic characters to help distinguish the species (Table 2). A spur is absent from *G. anomalus* (Fig. 10B) and *G. epedunculatus* (Figs. 2A, 25), and nearly absent in most specimens of *G. lateralis* (Fig. 4A) and

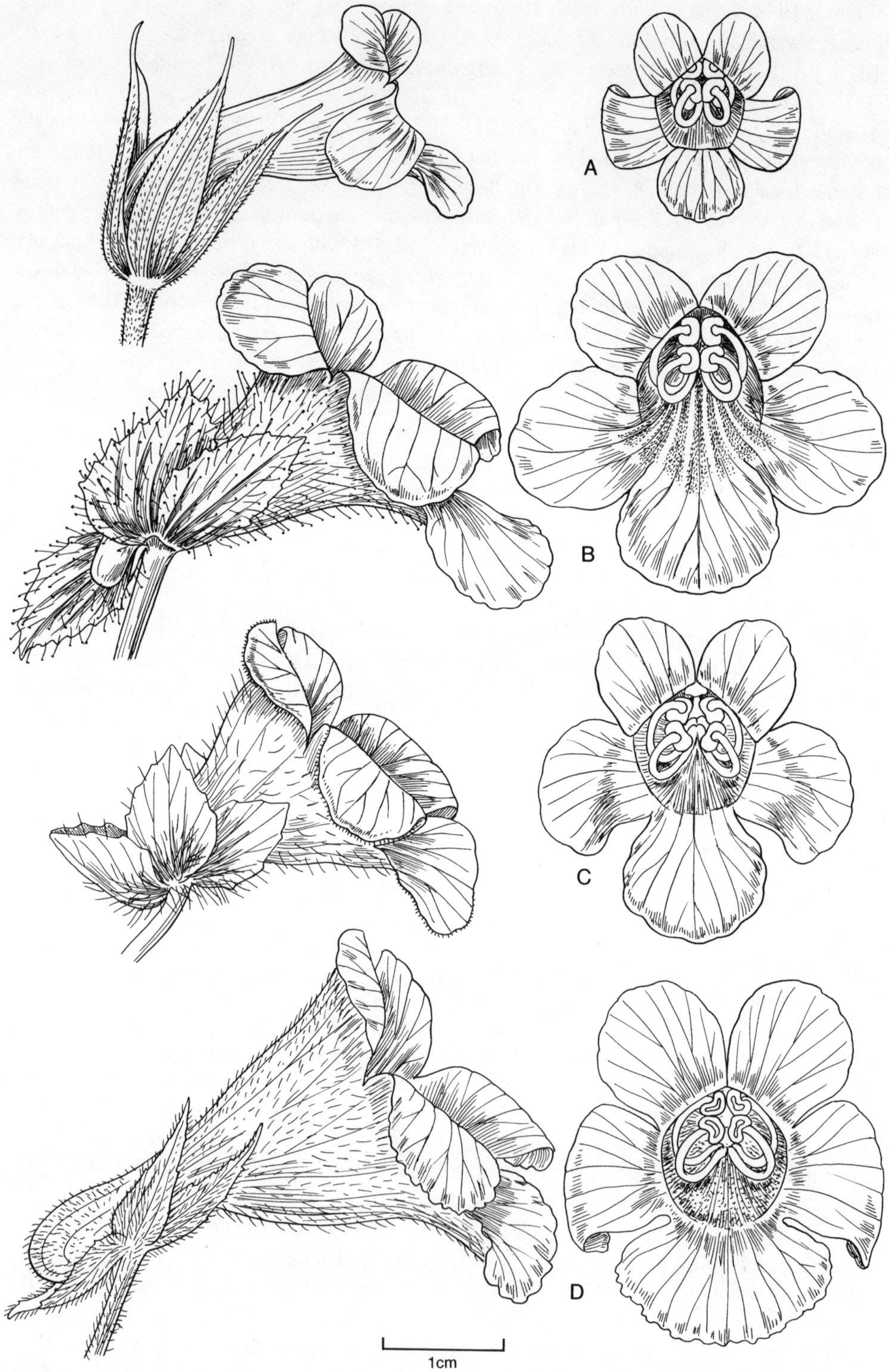

FIG. 2. Side and face views of flowers of *Gasteranthus*. A. *G. epedunculatus*. B. *G. adenocalyx*. C. *G. mutabilis*. D. *G. orientandinus*. (Based on: A, *Cuatrecasas 12866*; B, *Callejas et al. 6800*; C, *Jaramillo 5965*; D, *Camp E-1492*.)

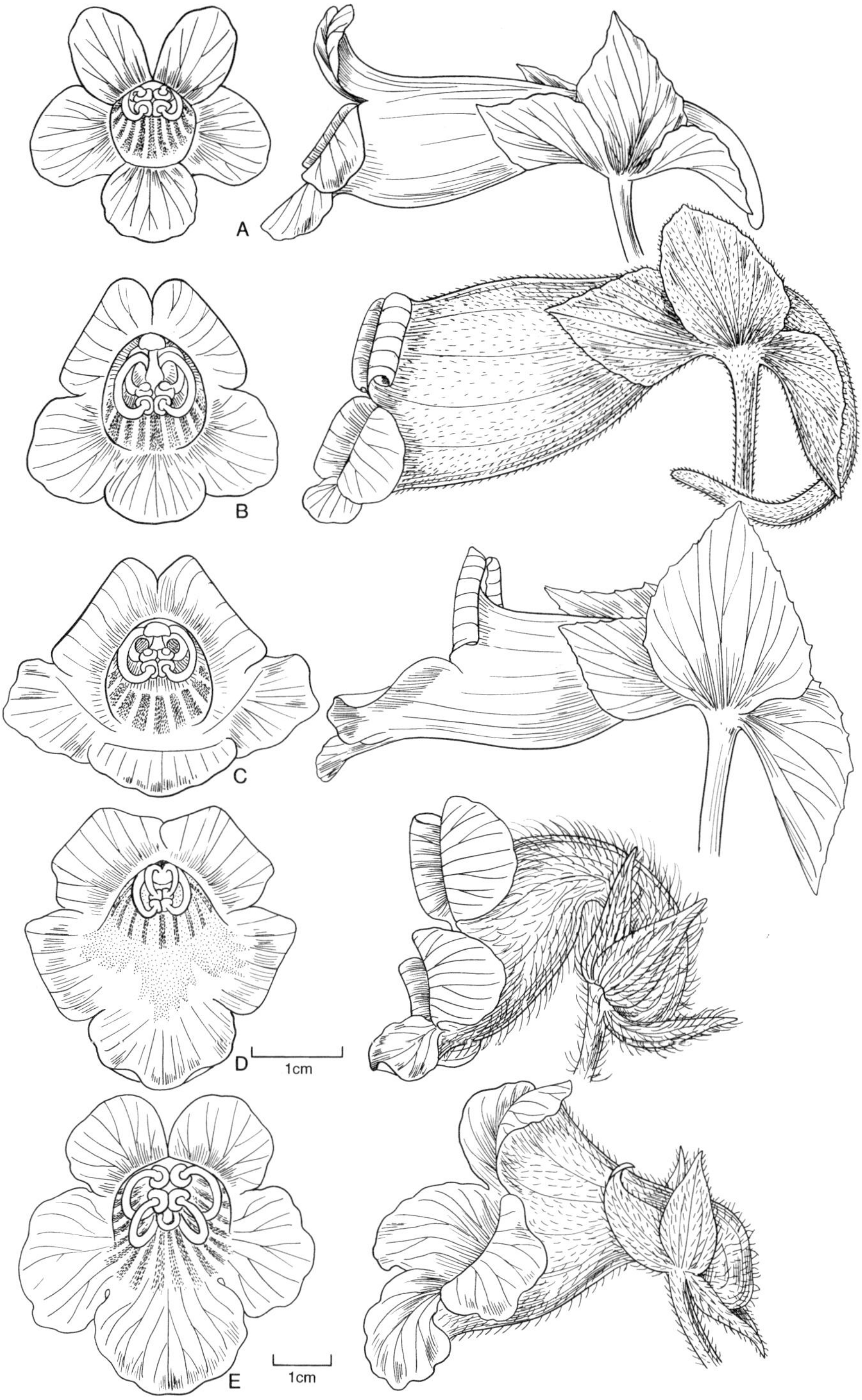

FIG. 3. Face and side views of flowers of *Gasteranthus*. A. *G. atratus*. B. *G. delphinioides*. C. *G. tenellus*. D. *G. dressleri*. E. *G. herbaceus*. (Based on: A, *Mexia 6175;* B, *Amaya & Kvist 377*, *Haught 4619*; C, *Clark 2815*; D, photo by J. Ertelt; E, *Sneidern 5464*, *Kvist 181*.)

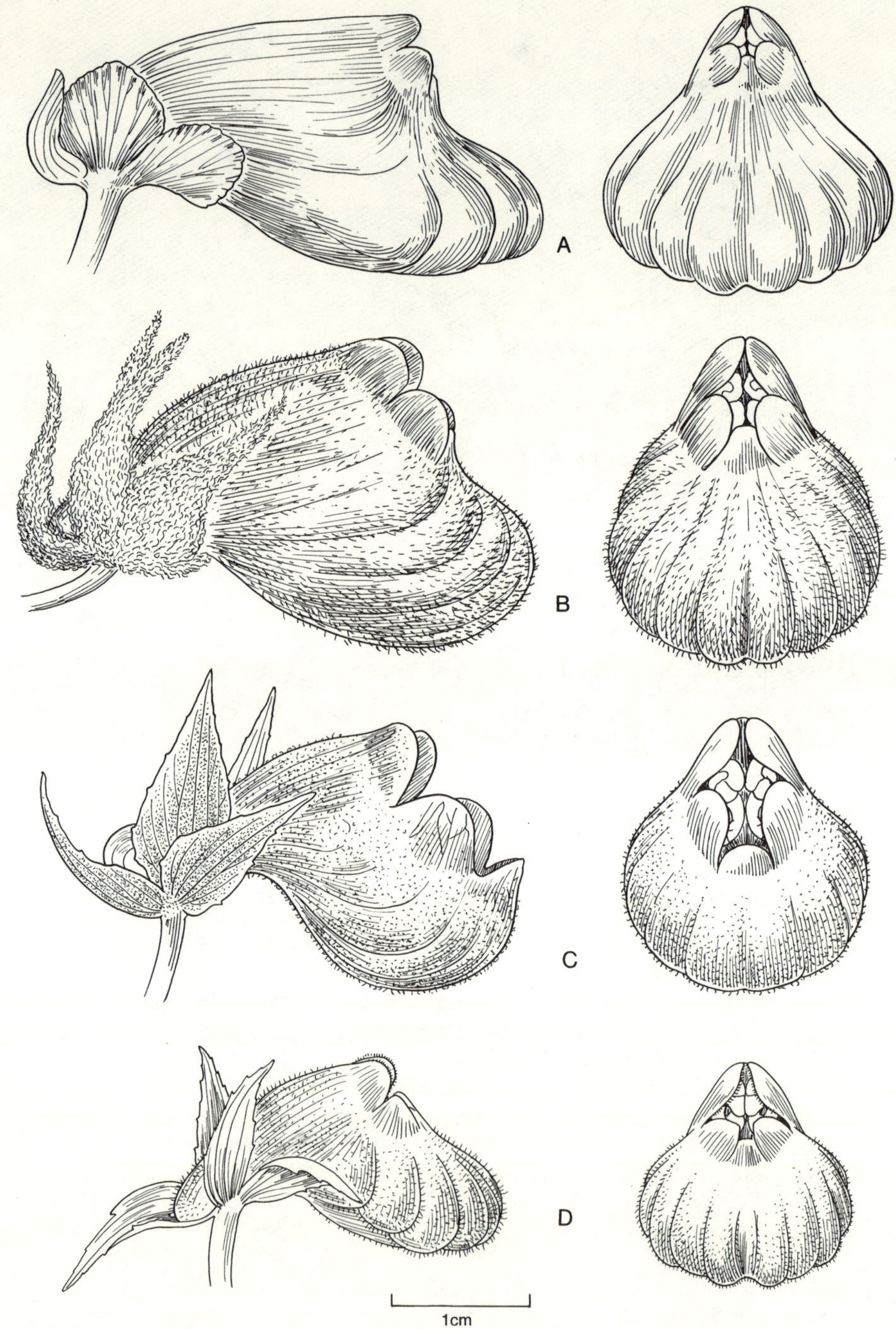

FIG. 4. Side and face views of flowers of *Gasteranthus*. A. *G. lateralis*. B. *G. macrocalyx*. C. *G. perennis*. D. *G. quitensis*. (Based on: A, *Mendoza-T. et al. 610*; B, *Dodson & Neill 15538*; C, *Holm-Nielsen 2899*; D, *Mendoza-T. et al. 525*, grown as USBRG 95-152.)

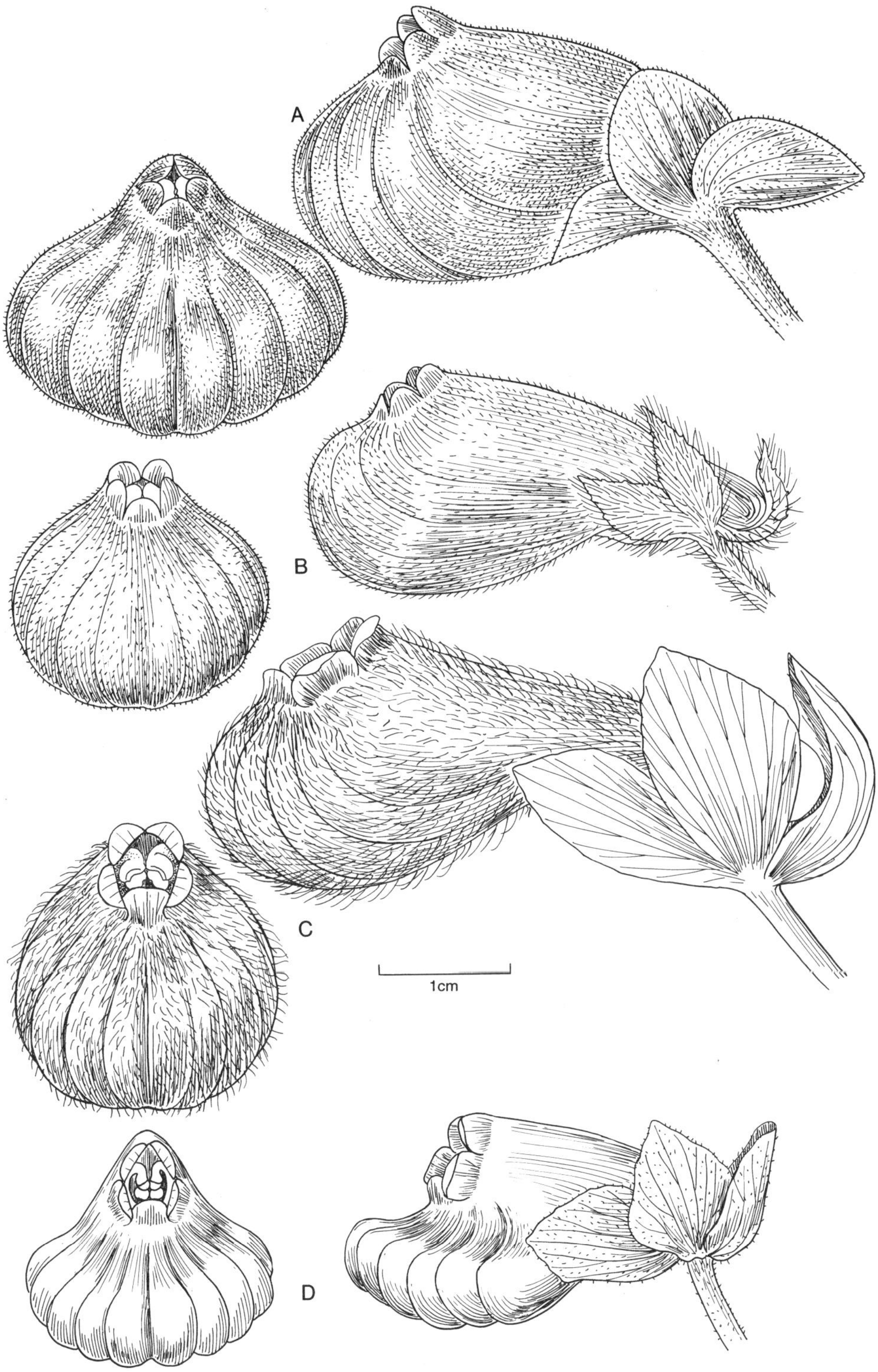

FIG. 5. Face and side views of flowers of *Gasteranthus*. A. *G. crispus*. B. *G. extinctus*. C. *G. ternatus*. D. *G. pansamalanus*. (Based on: A, *Mendoza-T. et al. 569*; B, *Dodson et al. 7117*; C, *Luteyn & Berg 14359*; D, *Standley 95056.*)

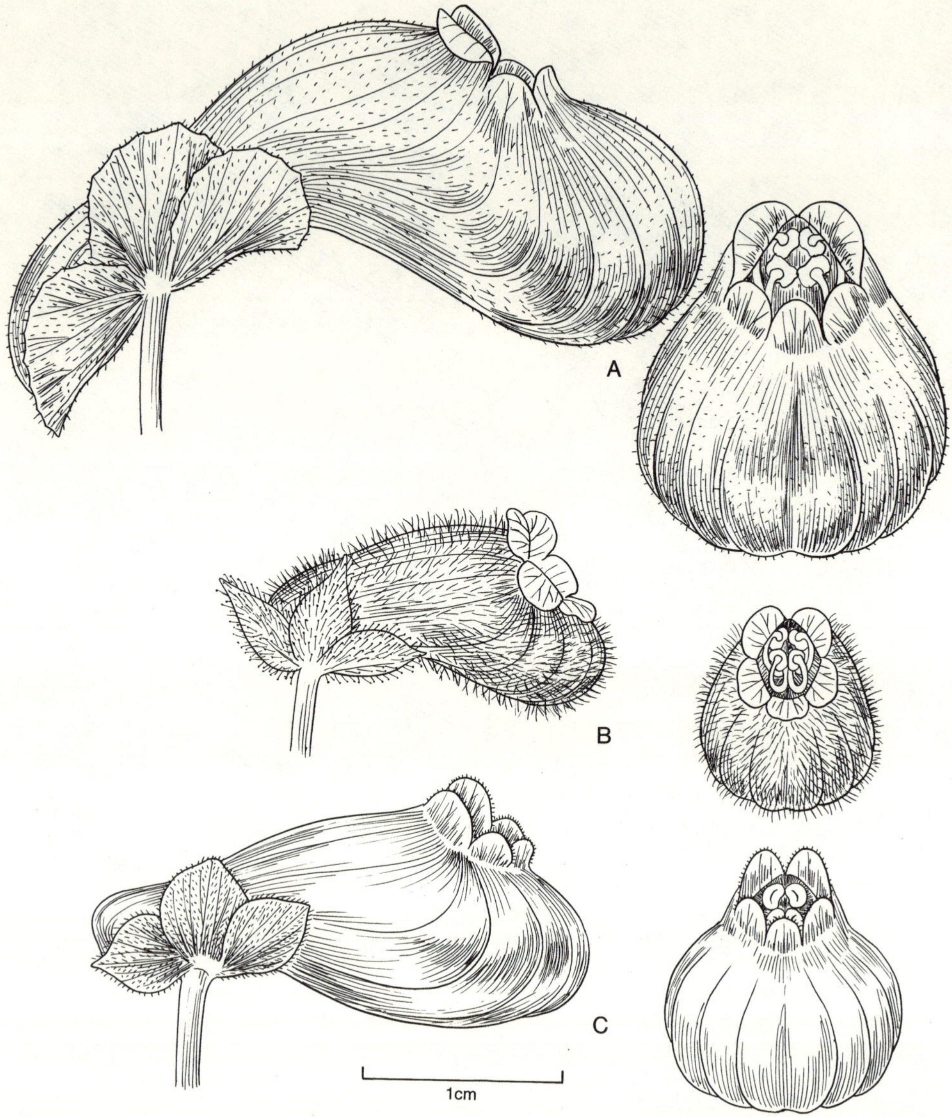

FIG. 6. Side and face views of flowers of *Gasteranthus*. A. *G. calcaratus* subsp. *calcaratus*. B. *G. calcaratus* subsp. *calceolus*. C. *G. calcaratus* subsp. *oncogastrus*. (Based on: A, *Ramos 1700*; B, *Kvist 128*; C, *Mendoza-T. et al. 600*.)

G. macrocalyx (Fig. 4B), although all these species are at least basally gibbous. It is noteworthy that two species, *G. imbricans* (Fig. 8C [spur partially hidden]) and *G. recurvatus* (Fig. 36), have spurs that are bent upward (Table 2).

Corolla colors may vary within populations of some species. For example, at the Bilsa Biological Station in northwestern Ecuador *G. calcaratus* subsp. *oncogastrus* usually has orange corollas, but some plants have yellow corollas. Plants of *G. crispus* mostly have orange corollas, but some occasionally have red corollas (J. Clark, pers. comm.).

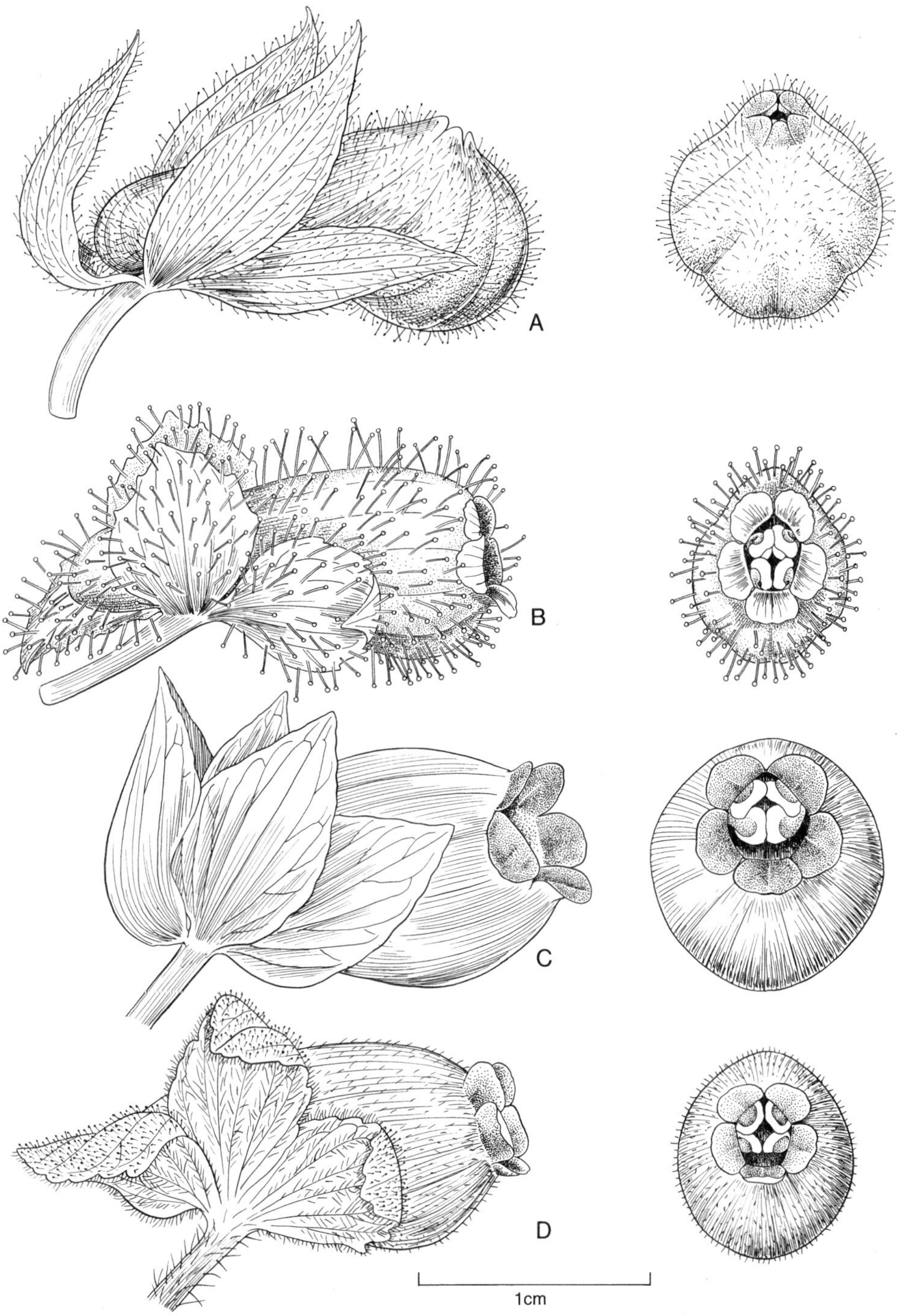

FIG. 7. Side and face views of flowers of *Gasteranthus*. A. *G. trifoliatus*. B. *G. imbaburensis*. C. *G. glaber* D. *G. bilsaensis*. (Based on: A, *Freiberg 96006*; B, *Freiberg 96238*; C, *Luteyn & Berg 14360*; D, *Clark et al. 1654*.)

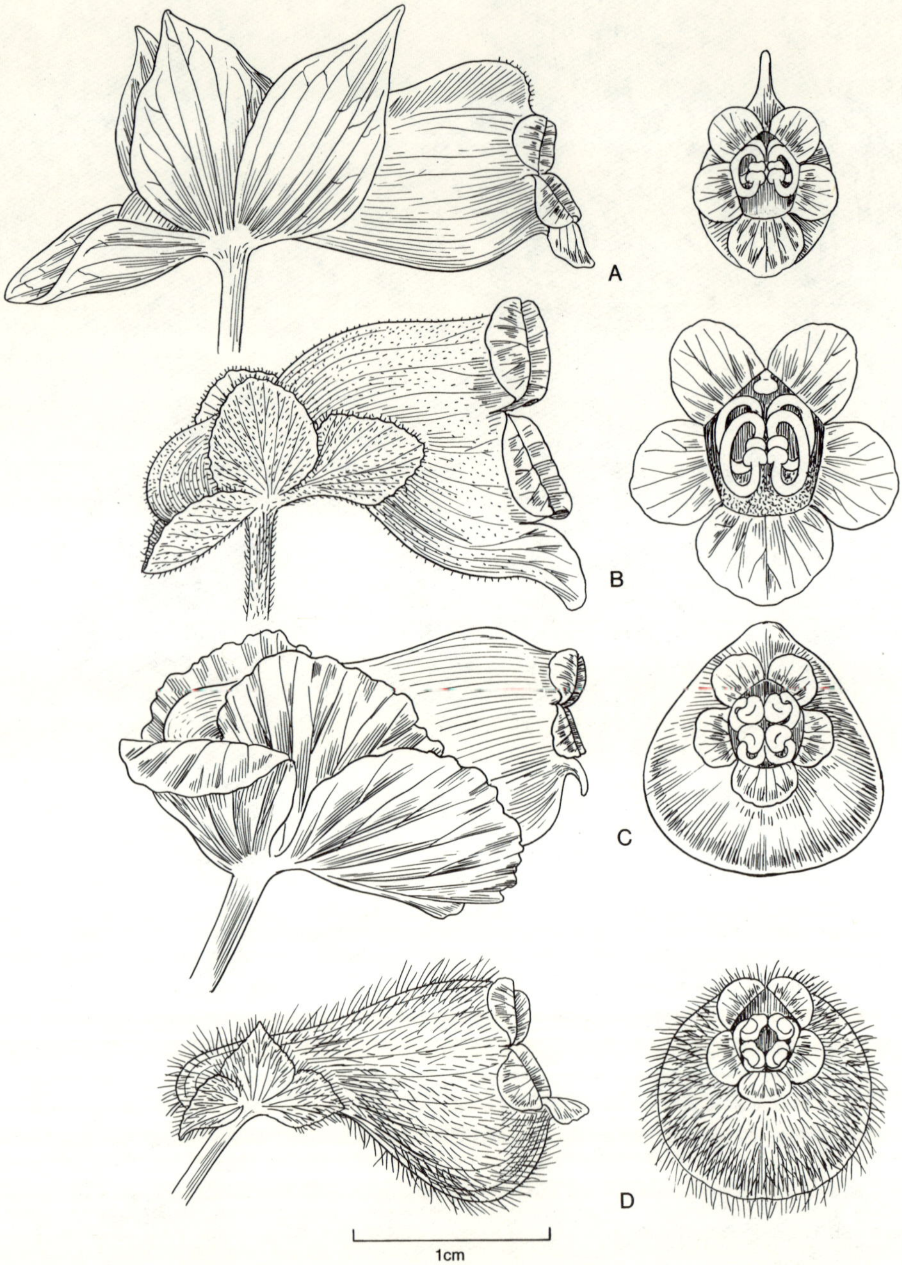

FIG. 8. Side and face views of flowers of *Gasteranthus*. A. *G. carinatus*. B. *G. columbianus*. C. *G. imbricans*. D. *G. timidus*. (Based on: A, *Dodson et al. 7515*; B, *Benavides 9621*; C, *Gómez 20321*; D, *Harling & Andersson 18907*.)

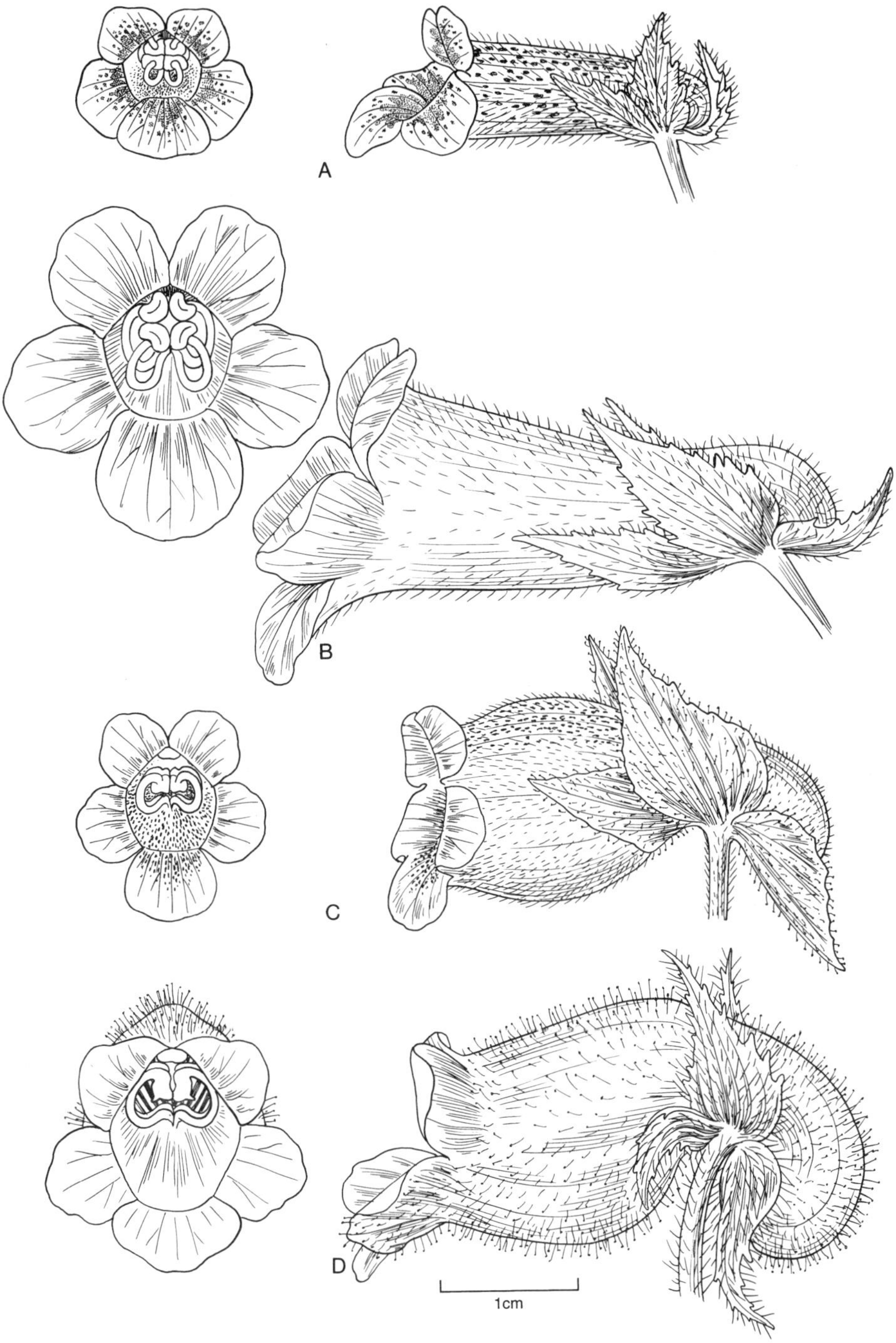

FIG. 9. Face and side views of flowers of *Gasteranthus*. A. *G. wendlandianus*. B. *G. leopardus*. C. *G. acropodus*. D. *G. villosus*. (Based on: A, *Lugo 6090*; B, *Schwerdtfeger 46*; C, *Skog & Linett 5537*; D, *Mendoza-T. et al. 594*, grown as USBRG 95-168.)

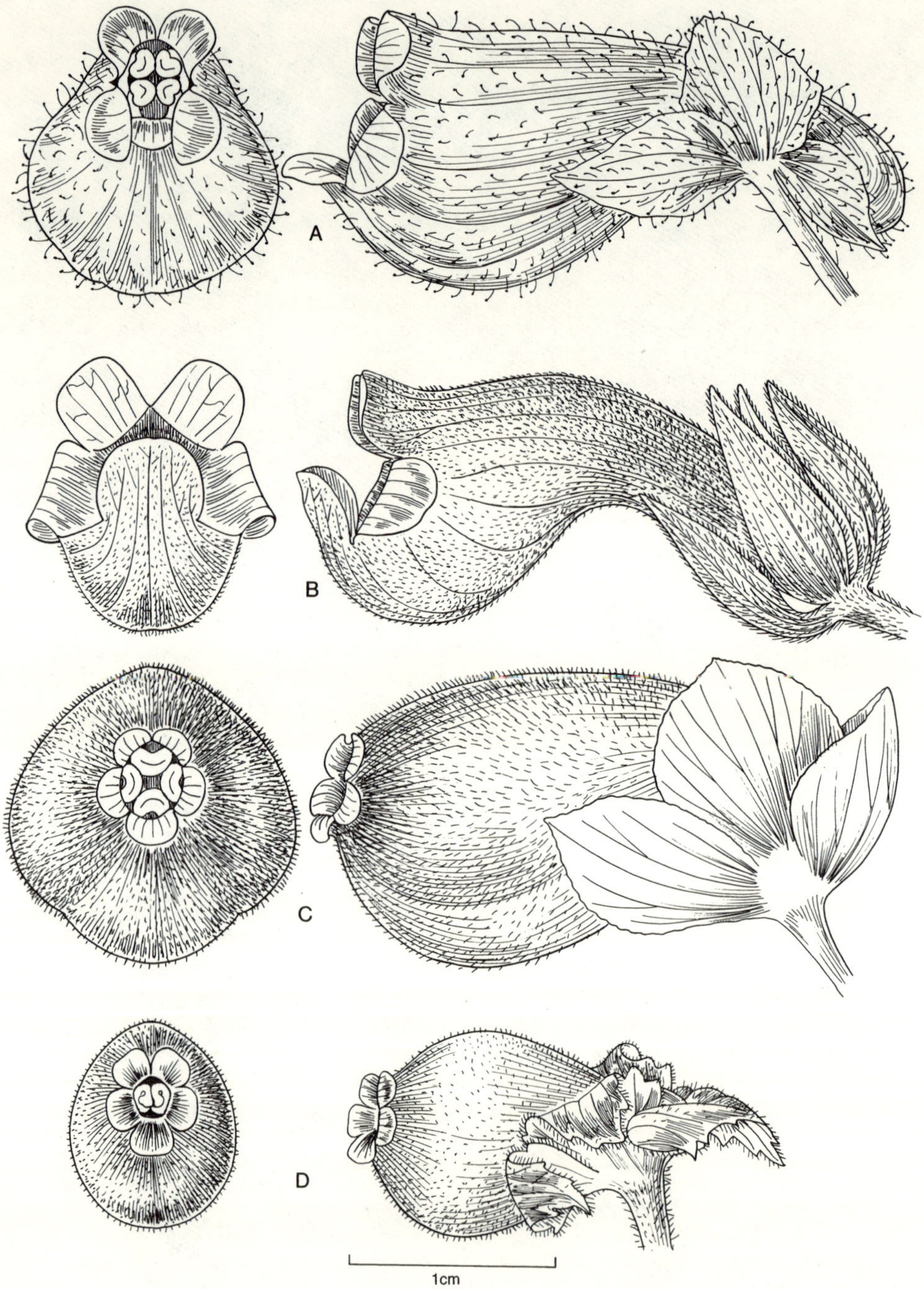

FIG. 10. Face and side views of flowers of *Gasteranthus*. A. *G. osaensis*. B. *G. anomalus*. C. *G. recurvatus*. D. *G. corallinus*. (Based on: A, *Allen 5533*; B, *Ramírez & López 324*; C, *Devia 594*, *Cuatrecasas 23978*; D, *Jaramillo & Coello 2722*, *Clark 2414*.)

Androecium and nectary. Gasteranthus, like most other New World members of Gesneriaceae, has four stamens with coherent, protandrous anthers (in the corollas in Figs. 2–9). In the later gynoecial stage of flowering the filaments tend to coil and pull the stamens back toward the base of the flower, but even in the earlier male stage of flowering the anthers never extend beyond the corolla throat, and are mostly located well inside the throat.

The filaments are curved (Figs. 2, 3, 9), so that although the filaments never extend beyond the corolla tube, their length often is greater than the distance from the base of the corolla to the corolla mouth. The filaments are mostly glabrous but some may be puberulent (see *G. leopardus*, Fig. 30), and the fifth stamen is sometimes present but reduced to a staminode (see *G. corallinus*, Fig. 21D). *Gasteranthus anomalus* is unusual in having filaments adnate to the corolla tube for most of their length, similar to *Cremosperma* (Table 1). All other species of *Gasteranthus* have filaments that are adnate only basally.

The nectaries vary considerably in *Gasteranthus*, but the most common type consists of two dorsal glands extending backward a short distance into the spur (see *G. leopardus*, Fig. 30F; *G. orientandinus*, Fig. 32F; and *G. recurvatus*, Fig. 36F). Others are more irregular but dorsal, or surround the ovary partly or entirely and are thickest or highest dorsally (see *G. corallinus*, Fig. 21F; *G. epedunculatus*, Fig. 25F; and *G. osaensis*, Fig. 33F). These latter nectaries thus approach the annular nectaries found in *Besleria* from which *Gasteranthus* may have been derived. In some species, e.g., *G. bilsaensis* (Fig. 17F), *G. leopardus* (Fig. 30F), *G. villosus* (Fig. 38F), etc., nectaries are pubescent but are glabrous in others. Nectarial pubescence is unusual and has not been mentioned by earlier authors in discussions of floral structures (Wilson 1974; Wiehler 1983). Nectaries also seem to vary within some species, as noted previously in some species of *Columnea* (Kvist & Skog 1993; Smith 1994) and *Kohleria* (Kvist & Skog 1992).

Gynoecium. These structures appear to provide few characters particularly useful for specific delimitation in *Gasteranthus*. The ovary as well as the style are glabrous or have a puberulent to velutinous indumentum, but specimens of the individual species tend to vary in this character.

The stigma varies from stomatomorphic (*G. orientandinus*, Fig. 32F) in most collections to weakly bilobed (*G. osaensis*, Fig. 33F). The distribution of stigma shapes varies among and within many species. The stigma, like the stamens, never extends beyond the throat of the corolla.

Fruit and seeds. In all species in which fruits have been observed, the fruit is a laterally compressed semi-fleshy to fleshy capsule that splits loculicidally with two reflexing valves (Fig. 11A), and sometimes secondarily septicidally with four valves (Fig. 12I), exposing a mass of seeds (see *G. columbianus*, Fig. 13C). Such fruits are common for many understory New World members of Gesneriaceae, but little is known concerning seed dispersal (Kvist & Skog 1992, 1996). Rain-wash in the very wet habitats of *Gasteranthus* and the accidental sticking of the tiny seeds to passing animals may contribute to seed dispersal. The disjunct distribution of *G. pansamalanus* (Fig. 34) suggests that long-distance dispersal may occasionally happen, possibly by birds eating the (berry-like) capsules before they open, or feeding on the seeds of open capsules. The fruit, however, remains surrounded by the persistent calyx, to some extent obscuring the fruit. Each capsule contains several hundred tiny seeds (usually maximally 0.3 mm long). The seeds are mostly ellipsoid or more irregularly shaped, and the surfaces are rather irregularly striate. Some species

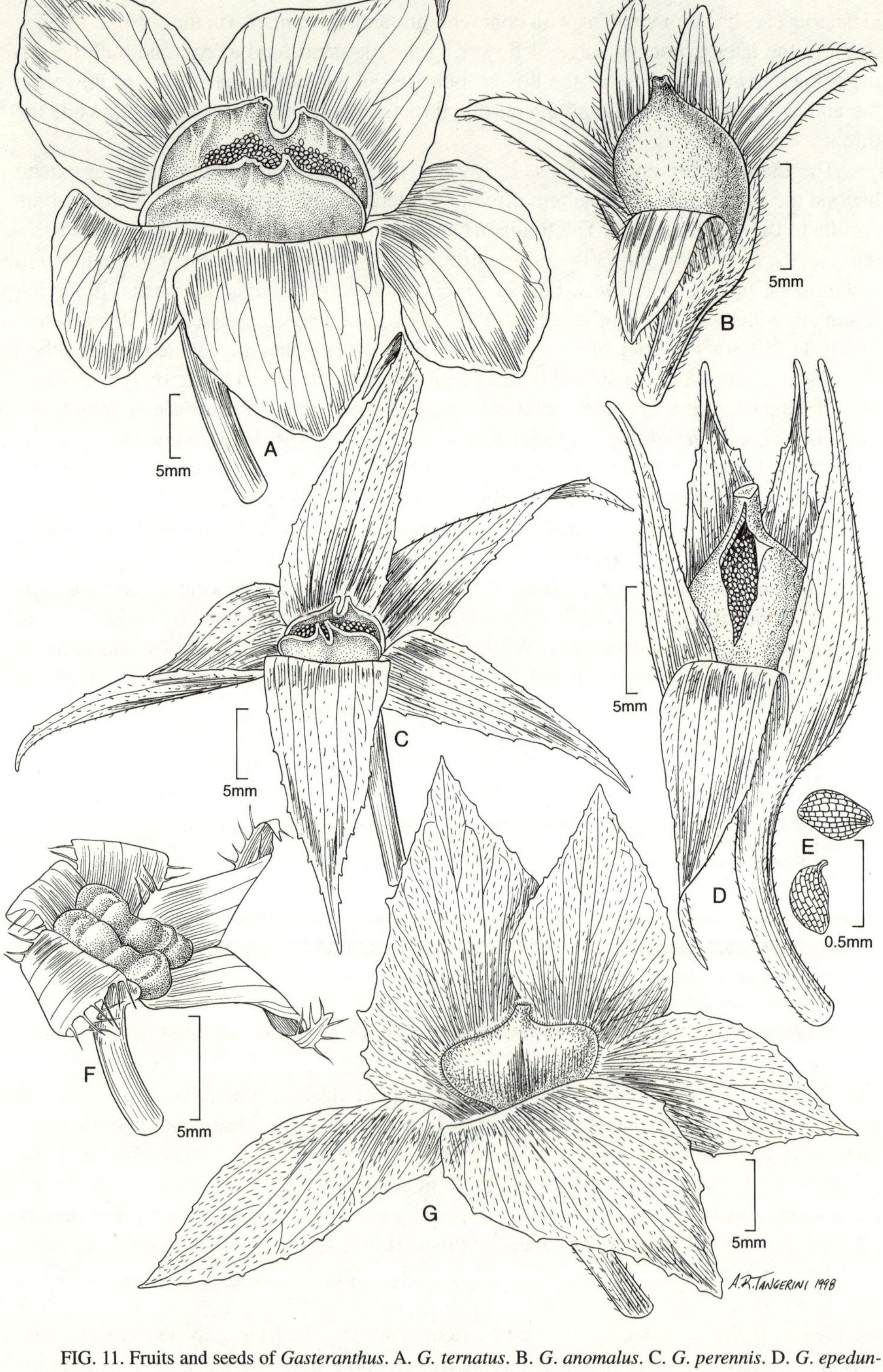

FIG. 11. Fruits and seeds of *Gasteranthus*. A. *G. ternatus*. B. *G. anomalus*. C. *G. perennis*. D. *G. epedunculatus*. E. *G. epedunculatus* (seeds). F. *G. quitensis*. G. *G. delphinioides*. (Based on: A, *Mendoza-T. et al. 623*; B, *Ramírez & López 324*; C, *Holm-Nielsen et al. 2899*; D, E, *Cuatrecasas 12866*; F, living material of *Mendoza-T. et al. 525*, grown as USBRG 95-152; G, *Amaya & Kvist 415*.)

FIG. 12. Fruits and seeds of *Gasteranthus*. A. *G. pansamalanus*. B. *G. calcaratus* subsp. *calceolus*. C. *G. calcaratus* subsp. *calceolus* (immature fruit). D. *G. calcaratus* subsp. *calceolus* (seeds). E. *G. lateralis*. F. *G. crispus*. G. *G. calcaratus* subsp. *calcaratus*. H. *G. calcaratus* subsp. *calcaratus* (seeds). I. *G. calcaratus* subsp. *oncogastrus*. J. *G. calcaratus* subsp. *oncogastrus* (seeds). (Based on: A, *Restrepo 498*; B, C, D, *Boeke 2218*; E, *Jaramillo 7875*; F, *Mendoza-T. et al. 569*; G, H, *Fraume & Gallego 218*; I, J, *Bass et al. 108*.)

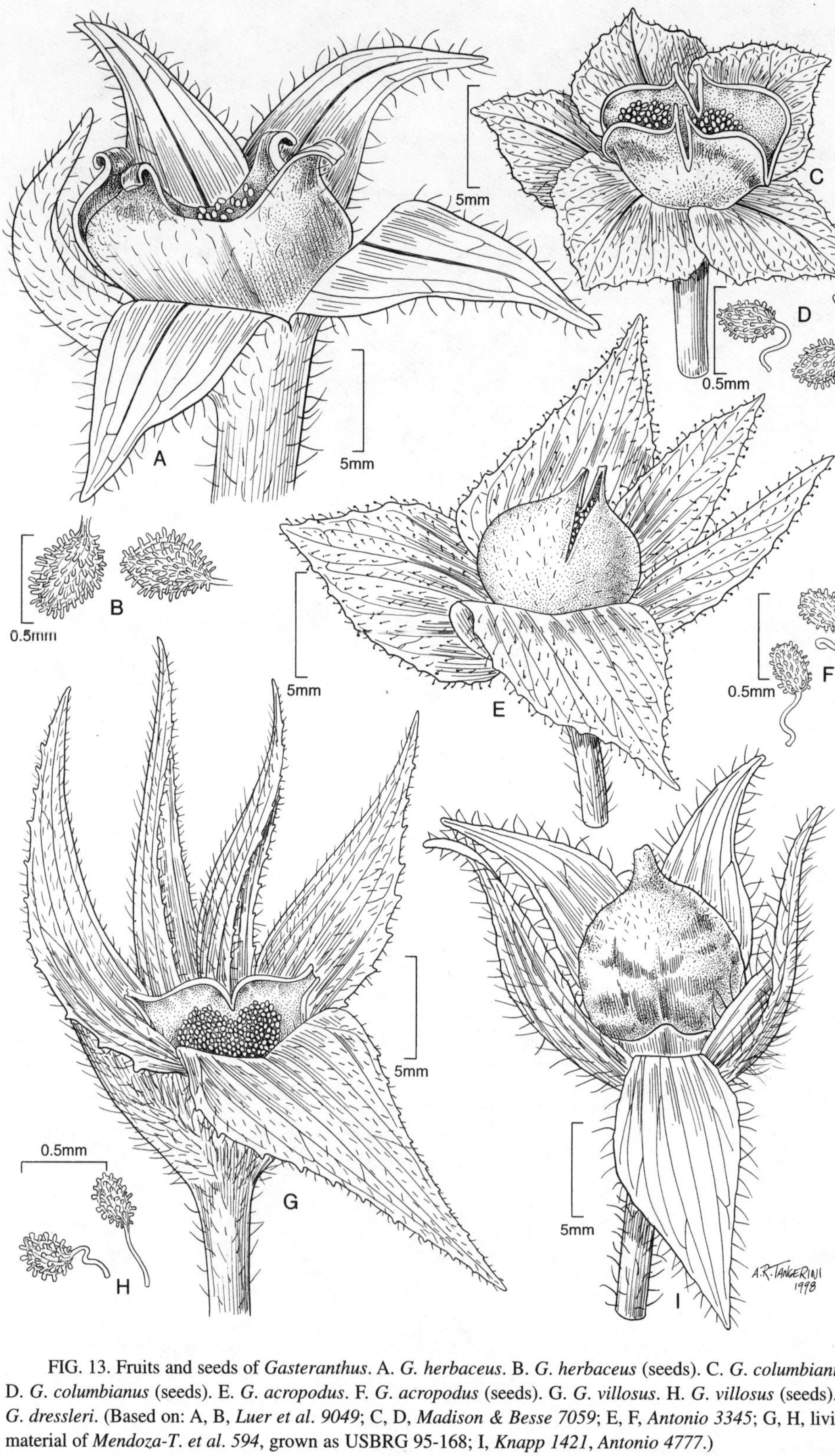

FIG. 13. Fruits and seeds of *Gasteranthus*. A. *G. herbaceus*. B. *G. herbaceus* (seeds). C. *G. columbianus*. D. *G. columbianus* (seeds). E. *G. acropodus*. F. *G. acropodus* (seeds). G. *G. villosus*. H. *G. villosus* (seeds). I. *G. dressleri*. (Based on: A, B, *Luer et al. 9049*; C, D, *Madison & Besse 7059*; E, F, *Antonio 3345*; G, H, living material of *Mendoza-T. et al. 594*, grown as USBRG 95-168; I, *Knapp 1421*, *Antonio 4777*.)

have seeds with papillate projections (see *G. calcaratus* subsp. *calceolus*, Fig. 12D; *G. calcaratus* subsp. *calcaratus*, Fig. 12H; *G. herbaceus*, Fig. 13B; *G. columbianus*, Fig. 13D; *G. acropodus*, Fig. 13F; *G. villosus*, Fig. 13H; *G. adenocalyx*, Fig. 16H; *G. orientandinus*, Fig. 32H; and *G. recurvatus*, Fig. 36H); however, seeds have been seen only from fewer than half of the species, making it difficult to appraise the taxonomic value of seed characters. Beaufort-Murphy (1983) depicted SEM-photos of seeds from *G. acropodus*, *G. atratus*, and *G. wendlandianus*. The last species (see Fig. 39F) has less conspicuous and much more regular and rounded striations than the former two species, and of these three species only *G. acropodus* (Fig. 13F) has papillate projections on the surface.

POLLINATION

Nearly half of the species of *Gasteranthus* have the whitish, funnelform to campanulate, or ventricose corollas with a conspicuous spur (Table 2; Figs. 2, 3, 9), which conform to the so-called "euglossophilous" flowers discussed by Wiehler (1978, 1983). Unfortunately, no field observations have so far substantiated that euglossine bees, in fact, predominantly pollinate those species. Nearly all species with "euglossophilous" flowers are either relatively rare or very locally distributed (see examples in Figs. 15, 24, 28). Exceptions are the large-flowered *G. delphinioides* (Fig. 24), which is very common in the Colombian department of Chocó (pers. obs.), and the widely dispersed small-flowered *G. wendlandianus*, found in montane forests from Costa Rica to Bolivia (Fig. 40). Species of *Gasteranthus* apparently rarely hybridize in natural habitats, and in the case of the presumed "euglossophilous" species of *Gasteranthus* this is not surprising. The species are mostly set apart by having different sizes and shapes of corollas and spurs (Table 2). In addition, the few species that have relatively similar corollas never seem to be sympatric (see the corolla shapes in Fig. 3 and the distributions of the same species in Figs. 24 and 28, also Fig. 2 and Fig. 15, and Fig. 9 and Figs. 15 and 40). The only case of what appears to be hybridization in the genus with the resulting intermediate progeny is between two "euglossophilous" species that have relatively similarly shaped but differently sized corollas. These are *G. wendlandianus* (Fig. 9A), with corollas ca. 2 cm long, and *G. leopardus* (Fig. 9B), with corollas to ca. 4 cm long. Nearly the entire continuum of corolla sizes between these two species can be found in the upper montane forest of La Planada in the Colombian department of Nariño. This may suggest that some pollinator has transferred pollen between the differently sized flowers. The "intermediate" specimens are herein referred to *G. leopardus* (see that species, no. 20).

The remainder of the species of *Gasteranthus* has relatively uniformly red, orange, or yellow corollas, and many of them conform to the "ornithophilous" urceolate (or hypocyrtoid) corolla type referred to by Wiehler (1978) (Figs. 4, 5, 6, 10), although others have funnelform and ventricose corollas (Table 2; Fig. 8). Urceolate corollas are inflated and usually have a conspicuous ventral pouch, the throat is constricted, and the limb is small. In more than half of the urceolate species the inflated pouch protrudes beyond the corolla throat (Figs. 2, 5, 6), implying that the distance from the base of the corolla to the tip of the pouch is considerably longer than the distance from the base to the corolla throat. There are reports of recent observations documenting that hummingbirds do, in fact, visit "ornithophilous" species of *Gasteranthus*. At the La Planada Reserve on the western Andean slopes in the Colombian department of Nariño, Gary Stiles and Marisol Amaya M. (pers. comm.) studied hummingbird pollination of plants of Gesneriaceae in montane forests. There the hummingbird *Aglaiocercus coelestis* was seen to visit three species of *Gasteranthus*, viz., *G. pansamalanus*, *G. corallinus*, and *G. columbianus*. That the same hummingbird may visit and possibly pollinate several species of *Gasteranthus* is not surprising, considering that many of the "ornithophilous" species (in contrast to the "euglossophilous" species) of *Gasteranthus* have fairly similar corollas (Table 2). *Gasteranthus pansamalanus* has a protruding pouch (Fig. 5D), *G. corallinus* has a pouch that is not protruding (Fig. 10D), and *G. columbianus* has a funnelform corolla without a pouch (Fig. 8B). However, the pouch may not influence the pollination directly, as the distance from the throat of the corolla to the dorsal nectary entering the spur at the base is within the same range in all the above-mentioned three species. This may allow the same birds to visit and feed on them, and suggests that the different types of pouches do not provide interspecific reproductive isolation. The pouch may alternatively help to attract the pollinators by making the flowers more conspicuous. It may also reduce the risk of self-pollination. In the female flowering stage, the stamens tend to be located at the bottom of the pouch, well below the receptive stigma approaching the corolla throat (although not in the distal part of pouches protruding beyond the throat of the corollas; stamens never extend beyond the throat). Several "ornithophilous" species of *Gasteranthus*, particularly *G. corallinus*, *G. calcaratus*, *G. quitensis*, and *G. pansamalanus*, are common and widespread, and thus often sympatric. Although they may have the same hummingbird visitors, hybridization seems to be rare, suggesting the presence of other reproductive barriers.

Ute Heppner (pers. comm.) studied pollination of plants of Gesneriaceae in a lowland Amazon rain forest at the Jatún Sacha Biological Station in the Ecuadorian province of Napo. She observed the hummingbird *Phaethornis aff. hispidus* visit flowers of *G. corallinus* (Fig. 21); however, she has also observed carpenter bees and euglossine bees visiting the flowers of the same species. She suspects (pers. comm.) that both hummingbirds and bees may pollinate *G. corallinus*; thus, the division into "ornithophilous" and "euglossophilous" species, based solely on pollination-syndromes rather than field-observations, is artificial.

HABITATS AND DISTRIBUTION

Plants of *Gasteranthus* are usually found in forests, and frequently in shaded and permanently humid places, e.g., in ravines with small streams or close to waterfalls. Many species are found in cloud forests, where they may be common in the understory vegetation. The same species may, however, also occur in forests at lower elevations than the cloud forests, where they are then restricted to the above-mentioned permanently humid places. For example, the western Ecuadorian species *G. crispus* is adapted to low-elevation cloud forests between 500 and 800 m elevation, and *G. lateralis* is adapted to high-elevation cloud forest above 1,800 m elevation; however, the two species may also occur locally in ravines down to 100 m and 900 m elevation, respectively (Fig. 14). A few species are less restricted to particular habitats and consequently more widespread and common. In western Ecuador (Fig. 1) this is particularly the case in *G. calcaratus* subsp. *oncogastrus* (Fig. 18) and *G. quitensis* (Fig. 35). The latter is locally common in remnant vegetation along roadsides in recently deforested low montane forests. Here *G. quitensis* tends to grow protected in the shadows of larger herbs, shrubs, or small trees rather than exposed to direct sun. *Gasteranthus calcaratus* subsp. *oncogastrus* was probably common

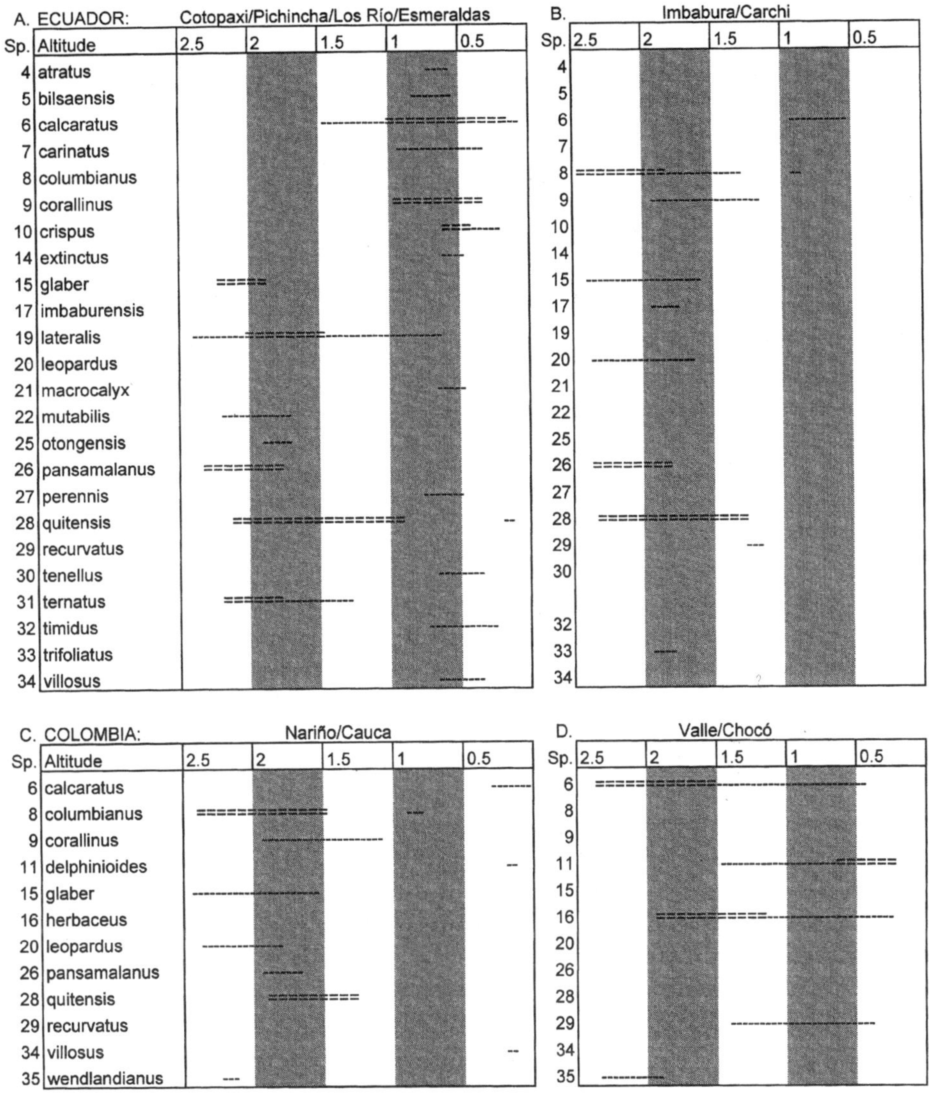

FIG. 14. Altitudinal distribution of *Gasteranthus* in western Ecuador and Colombia, ranging from provinces in central Ecuador in the south (A and B) to departments in northwestern Colombia (C and D). Elevations range from sea level at the right to 2,500 m at the left. Single and double lines indicate uncommon and common species, respectively. A. Ecuador: Cotopaxi, Pichincha, Los Ríos, and Esmeraldas Provinces. B. Ecuador: Imbabura and Carchi Provinces. C. Colombia: Nariño and Cauca Departments. D. Colombia: Valle del Cauca and Chocó Departments.

near the coast in the now mostly destroyed seasonal forests from sea level to 900 m or more.

The 35 species of *Gasteranthus* range mostly from Costa Rica and Panama through western South America south to Bolivia, with the center of diversity in western Ecuador and adjacent western Colombia (Fig. 1). Ecuador has 26 species (including two subspecies) of *Gasteranthus*; 16 are endemic. Colombia has 15 species (including three subspecies); four are endemic (Appendix). In Peru only three widespread species, *G. calcaratus* (Fig. 18), *G. corallinus* (Fig. 22), and *G. wendlandianus* (Fig. 40), occur, and only the last species reaches Bolivia. To the north five species each are found in Panama and Costa Rica. Four species are endemic to Central America, viz., *G. acropodus* (Fig. 15), *G. dressleri* (Fig. 28), and *G. imbricans* (Fig. 20), all mostly found near the more humid Caribbean coast, and *G. osaensis* (Fig. 20) on the Pacific coast but restricted to southeastern Costa Rica, mainly on the Osa Peninsula. The remaining two species in Costa Rica and Panama are the widespread *G. wendlandianus* (Fig. 40) and *G. delphinioides* (Fig. 24), both of which are also common in northwestern Colombia. One otherwise mainly South American species, *G. pansamalanus*, also occurs in Guatemala and nearby Mexico (Fig. 34).

In both Ecuador and Colombia *Gasteranthus* is more frequent in the west (Figs. 1, 14), but some common species have wider distributions, and there even are a few endemic species in other parts of the two countries. Six species occur in eastern Ecuador and eastern Colombia, defined here as the eastern Andean slopes and the lowlands draining to the Amazon and Orinoco Rivers. Four of these species are widespread, viz., *G. calcaratus* (Fig. 18), *G. corallinus* (Fig. 22), *G. pansamalanus* (Fig. 34), and *G. wendlandianus*, (Fig. 40), whereas *G. orientandinus* (Fig. 15) and *G. epedunculatus* (Fig. 15) are rare endemics of southwestern Ecuador and northeastern Colombia, respectively. In northwestern Colombia *Gasteranthus* is relatively well represented. The department of Antioquia has six species including the two rare endemics, *G. anomalus* (Fig. 19) and *G. adenocalyx* (Fig. 15). *Gasteranthus herbaceus* (Fig. 28) is also restricted to northwestern Colombia but has a wider distribution. *Gasteranthus recurvatus* (Fig. 19) occurs in western Colombia in the Department of Valle del Cauca. It has apparently been collected once (but without any corollas) in the province of Carchi of northwestern Ecuador, and can thus be expected from the intervening Colombian departments of Cauca and Nariño.

The distribution patterns of the 25 species of *Gasteranthus* found in western Ecuador and adjacent southwestern Colombia cannot be interpreted without proper consideration of their altitudinal ranges and habitats. Particularly in Ecuador, most species are confined either to forests below 1,000 m elevation or to forests at higher elevations (Fig. 14). Much of western Ecuador usually has two layers of clouds that shroud forests at different altitudes. These layers of clouds thus impinge on distinct low-elevation and high-elevation cloud forests; the former between ca. 600 and 800 m elevation and the latter above 1,800 m elevation (see discussion by R. Foster in Parker & Carr 1992).

In western Ecuador low-elevation cloud forests occur both along the Andean slopes and in a range of up to 900 m high coastal hills separated from the Andean slopes by a nearly 100 km wide but maximally 300 m high plateau. Six species of *Gasteranthus* occur in the coastal hills. Three of them, *G. bilsaensis* (Fig. 22), *G. tenellus* (Fig. 24), and *G. villosus* (Fig. 15), are endemic to coastal hill cloud forests and adjacent wet forests (including extreme southwestern Colombia near Tumaco where the last species has been collected). Apparently, the most extensive area of remaining coastal cloud forest is found in the Cordillera de Mache south of the town Esmeraldas, and has recently become known as the Bilsa Biological Station (Parker & Carr 1992). Here also, in addition to the abovementioned three newly described species, occur *G. crispus* (Fig. 23) and *G. calcaratus* subsp. *oncogastrus* (Fig. 18). The former is primarily adapted to low-elevation cloud forests but has a wider distribution along shaded creeks in western Ecuador (Dodson &

Gentry 1978; pers. obs.). The latter is more tolerant of varying humidity levels, and is widely distributed and common in western Ecuador.

Along the Andean slopes, the most well-known and studied low-elevation cloud forest is found on Centinela Ridge, a 600 m high western Andean front range on the border of Pichincha and Los Ríos provinces. According to Dodson and Gentry (1991), six species of *Gasteranthus* are endemic to Centinela Ridge. The present work demonstrates that one of these six species, *G. extinctus* (Fig. 23), never was collected elsewhere; however, *G. atratus* (Fig. 24) was also found on nearby Andean slopes east of Centinela Ridge. *Gasteranthus carinatus* (Fig. 19), *G. macrocalyx* (Fig. 29), and *G. perennis* (Fig. 35) have all been found in other low-elevation cloud forests, but only along the Andean slopes and at localities south of Centinela Ridge. The sixth species, *G. timidus* (Fig. 20), also occurs to the west in the coastal hills of Manabí province (south of Bilsa but apparently not present at Bilsa). North of Centinela Ridge, in western Ecuador as well as further northward through western Colombia, no species of *Gasteranthus* apparently have specifically adapted to low-elevation cloud forests (Fig. 14). Low-elevation cloud forests along the central and southern Andean slopes in western Ecuador thus have five resident endemic species of *Gasteranthus*. By adding *G. calcaratus* and *G. crispus* to the above-mentioned six (Centinela) species, at least eight species once occurred in these small and local forests.

The upper-elevation cloud forests are much more extensive and continuous than the low-elevation cloud forests. Two of the more widespread species of *Gasteranthus* are adapted to this habitat, viz., *G. wendlandianus*, found from Costa Rica to Bolivia (Fig. 40), and *G. pansamalanus*, distributed in Ecuador and Colombia, and disjunct in Guatemala and adjacent Mexico (Fig. 34). Most species of *Gasteranthus* adapted to the upper cloud forests have limited ranges. *Gasteranthus otongensis* is known only from the province of Cotopaxi (Fig. 23), *G. mutabilis* (Fig. 15) and *G. ternatus* only from Pichincha (Fig. 23), and *G. imbaburensis* (Fig. 20) and *G. trifoliatus* (Fig. 23) only from Imbabura; *G. lateralis* is known from Pichincha and Imbabura (Fig. 29). Three species, *G. columbianus* (Fig. 20), *G. leopardus* (Fig. 40), and *G. glaber* (Fig. 19), have a larger but still fairly limited range in the upper cloud forests from northwestern Ecuador to the southwestern Colombian department of Nariño. *Gasteranthus recurvatus* is so far known from Ecuador only from a questionable sterile specimen from the province of Carchi near the Colombian border (Fig. 19), and otherwise from the department of Valle del Cauca further to the north in Colombia. *Gasteranthus quitensis* (Fig. 35) has about the same distribution as *G. columbianus*, *G. leopardus*, and *G. glaber*, but occurs farther to the south in western Ecuador, and is also common in lower montane forests locally down to 700 m elevation (Fig. 14). *Gasteranthus corallinus* (Fig. 22) is also found in the upper cloud forests of northwestern Ecuador and adjacent Colombia, but has a much more extensive distribution throughout Ecuador in lowland forests and low montane forests primarily in the western Amazon Basin into Peru. It seems to be restricted to forests receiving at least 3,000 mm of annual precipitation. The three geographically separated subspecies of *G. calcaratus* (Fig. 18) mostly occur in different forest habitats. Subspecies *calceolus* from the eastern Andean slopes occurs in wet lowland rain forests but also in lower montane forests, and subsp. *calcaratus* from western Colombia occurs in upper cloud forests and upper montane forests. Subspecies *oncogastrus* from western Ecuador (and extreme southwestern Colombia) mostly occurs in moist lowland forests (1,500 to 3,000 mm annual precipitation), but occasionally also in wet lowland forests, in low-elevation cloud forests, and in lower montane forests.

CONSERVATION STATUS

Gasteranthus extinctus, here newly described, may already be extinct, and this may also be true for *G. orientandinus* and *G. epedunculatus*. Nearly half of the 35 species of *Gasteranthus* may be endangered or already extirpated from known localities. The extinction has resulted from forest destruction causing lower humidity and higher light levels detrimental to living plants of *Gasteranthus*. We have observed from growing several species in the greenhouses at the Smithsonian Institution that high humidity and low light levels are necessary. Particularly the deforestation of the low-elevation cloud forests in western Ecuador, from which 10 species are known, has been serious. Dodson and Gentry (1991) reported that six species of *Gasteranthus* were endemic to Centinela Ridge, and that nearly 100 species of other flowering plants may have become extinct following the complete deforestation of this 600 m high Andean front range. Only one of the six species of *Gasteranthus*, viz., *G. extinctus* (Fig. 23), is known only from Centinela Ridge, whereas *G. atratus* (Fig. 24), *G. carinatus* (Fig. 19), *G. macrocalyx* (Fig. 29), and *G. perennis* (Fig. 35) also occur in other low-elevation cloud forests along the Andes. Unfortunately, these forests now may also be destroyed, and it may be too late to preserve intact low-elevation cloud forest along the western Andean slopes, although some disturbed remnants of this vegetation probably still exist. The sixth Centinela species, *G. timidus*, has been found in a small cloud forest south of Centinela in Los Ríos province, and probably is also still found in the coastal hills of Manabí province (Fig. 20). Here little-explored (but somewhat disturbed) cloud forests remain in the Machalilla National Park and apparently also in the Colonche hills southwest of Machalilla; and *G. timidus* may thus survive in these areas. Further to the north in the Cordillera de Mache of Esmeraldas province exist extant populations of the four remaining endangered western Ecuadorian low-elevation cloud forest species, *G. bilsaensis* (Fig. 22), *G. crispus* (Fig. 23), *G. tenellus* (Fig. 24), and *G. villosus* (Fig. 15). These species will only survive if significant tracts of this still fairly extensive forested area are preserved. Fortunately, the Bilsa Biological Station (owned by the Jatún Sacha Foundation) now protects 3,000 ha of forests. In November, 1996, the Ecuadorian Park Service declared a 70,000 ha area surrounding the Bilsa Biological Station a National Ecological Reserve, probably ending the granting of timber concessions. An extant population of *G. bilsaensis* also exists ca. 40 km south of Bilsa in the small cloud forest at the 800 m high and isolated coastal hill Cerro Pato de Pájaro.

The survival of some upper cloud forest species may also be threatened in western Ecuador, particularly of *G. imbaburensis* (Fig. 20), *G. lateralis* (Fig. 29), *G. mutabilis* (Fig. 15), *G. otongensis* (Fig. 23), *G. ternatus* (Fig. 23), and *G. trifoliatus* (Fig. 23), which all have very limited ranges. Extant populations remain of these species, but apparently not within any large conservation units. *Gasteranthus lateralis*, *G. mutabilis*, and *G. ternatus* occur at the small, private "La Guajalito" Reserve, located at approx. 1,900 m elevation along the old Quito–Santo Domingo road. Here also occur *G. glaber*, *G. pansamalanus*, and *G. quitensis*, implying that this upper-elevation cloud forest with six species matches the richest low-elevation cloud forests in *Gasteranthus* diversity. Finally, *G. imbaburensis*, recently described from Los Cedros Biological Reserve, and *G. otongensis*, recently described from the Bosque Protector Otonga in Cotopaxi, may be currently protected.

Outside western Ecuador, it is more difficult to estimate the status of *Gasteranthus,* but several species probably are endangered or perhaps already extinct. *Gasteranthus orientandinus* (Figs. 15, 32) from southwestern Ecuador and *G. epedunculatus* (Figs. 15,

25) from northwestern Colombia may also be extinct. Both have not been collected for more than 50 years, and were originally collected within areas that have since been deforested. *Gasteranthus anomalus* (Fig. 19) and *G. adenocalyx* (Figs. 15, 16) are only known from a few very local (although more recent) collections in northwestern Colombia. *Gasteranthus dressleri* (Fig. 28) and *G. osaensis* (Figs. 20, 33) have equally small ranges in Panama and Costa Rica, respectively, but extant populations are known.

Summing up, throughout the range of *Gasteranthus* at least seven species may already be extinct, and the survival of more than 10 species may be seriously at risk. Only one of the possibly extinct species, *G. atratus*, is known to be in cultivation; four endangered species found at Bilsa recently were introduced in cultivation at the Smithsonian Institution. Kvist et al. (unpubl.) analyze and discuss the conservation status of all members of Gesneriaceae recorded below 1,000 m elevation in western Ecuador. Ten species are presumed to be extinct in this area and 26 species to be endangered, corresponding to more than a third of the 107 species of Gesneriaceae recorded from western Ecuador. We estimate (Kvist et al., unpubl.) 11 of the 15 species of *Gasteranthus* recorded below 1,000 m elevation in western Ecuador to be extinct or endangered, implying that the survival of this genus is even more at risk than that of other representatives of Gesneriaceae found in the region.

TAXONOMY

Gasteranthus Bentham, Pl. Hartweg. 233. 1846. *Besleria* subg. *Gasteranthus* (Bentham) Hanstein, Linnaea 34: 334. 1865. *Besleria* sect. *Gasteranthus* (Bentham) Bentham in Bentham & Hooker f., Gen. pl. 2(2): 1015. 1876.—TYPE: *Gasteranthus quitensis* Bentham.

Halphophyllum Mansfeld, Repert. Spec. Nov. Regni Veg. 41: 145. 1936.—TYPE: *Halphophyllum crispum* Mansfeld [=*Gasteranthus crispus* (Mansfeld) Wiehler].

Besleria sect. *Neobesleria* subsect. *Corallinae* C. V. Morton, Contr. U.S. Natl. Herb. 26: 414. 1939.—TYPE: *Besleria corallina* Fritsch [=*Gasteranthus corallinus* (Fritsch) Wiehler].

Besleria sect. *Neobesleria* subsect. *Imbricatae* C. V. Morton, Contr. U.S. Natl. Herb. 26: 415. 1939.—TYPE: *Besleria imbricans* J. D. Smith [=*Gasteranthus imbricans* (J. D. Smith) Wiehler].

Besleria sect. *Neobesleria* subsect. *Maculatae* C. V. Morton, Contr. U.S. Natl. Herb. 26: 415. 1939.—TYPE: *Besleria rupestris* C. V. Morton [=*Gasteranthus wendlandianus* (Hanstein) Wiehler].

Besleria sect. *Neobesleria* subsect. *Wendlandianae* C. V. Morton, Contr. U.S. Natl. Herb. 26: 415. 1939.—TYPE: *Besleria wendlandiana* Hanstein [=*Gasteranthus wendlandianus* (Hanstein) Wiehler].

Besleria sect. *Neobesleria* subsect. *Herbaceae* C. V. Morton, Contr. U.S. Natl. Herb. 26: 416. 1939.—TYPE: *Besleria herbacea* C. V. Morton [=*Gasteranthus herbaceus* (C. V. Morton) Wiehler].

Plants mostly herbs, more or less suffruticose basally, occasionally small shrubs; stems mostly erect, occasionally decumbent or appressed to the ground, mostly with an indumentum at least distally but usually glabrescent on older parts, sometimes glabrous, internodes terete or quadrangular. Leaves predominantly opposite, rarely ternate, or very rarely whorled, usually isophyllous to subisophyllous, occasionally conspicuously anisophyllous but only rarely strongly anisophyllous, blades most commonly elliptic and ovate, less commonly obovate, oblanceolate, rarely narrowly elliptic or narrowly oblanceolate, usually flat, sometimes rugose, base usually cuneate or acute, less commonly attenuate or obtuse, apex mostly acute, frequently acuminate, occasionally obtuse, margin mostly serrate, less commonly crenate or subentire, occasionally serrulate or dentate, adaxially more or less dark green, sometimes nearly black, mostly glabrous or appressed-pilose, sometimes muricate, papillate, or strigillose, abaxially lighter green than adaxially, sometimes purplish or darker at or along veins, most frequently pilose, villous, tomentose, arachnoid, or strigillose, indument denser along the veins, lateral veins 5–17 per side but most frequently 7–11 per side, stomata aggregated in conspicuous or inconspicuous clusters; petioles present, sometimes absent, indumentum mostly similar to that of lower leaf surface, particularly of the primary vein. Inflorescences varying from densely congested to open cymes, of (1–) 2–8 (–30) protandrous flowers, indumentum conspicuous or appressed or sometimes absent; peduncles usually much longer than petioles, rarely shorter or absent. Calyx mostly green or light green, occasionally yellow, reddish, or purplish, mostly with a conspicuous indumentum outside but very variable, sometimes glabrous, inside more frequently glabrous or with a less conspicuous indumentum, sometimes glandular-hairy on one or both surfaces, lobes nearly free to base or briefly connate, mostly subequal but often appearing more unequal due to varying lobe orientation, when unequal differing principally in size rather than shape, each of the lobes usually ovate or lanceolate, occasionally deltate or narrowly lanceolate, apex acute or acuminate, occasionally obtuse, dorsal lobe often keeled, lobes often larger in fruiting stage than in flowering stage, margin mostly entire to subentire but also serrate, crenate, dentate, erose, and toothed, dorsal lobe usually directed backward below and surrounding spur, if the lobe is longer than the spur it may continue straight backward or secondarily be directed upward, ventral lobes mostly directed forward and appressed to corolla, sometimes upward, lateral lobes mostly directed upward at each side of the base of the corolla, sometimes more or less forward appressed to corolla. Corolla varying from urceolate with a narrow throat and a small limb and a large ventral pouch to funnelform-campanulate with a large and wide limb, between these extremes occur ventricose, subventricose, and funnelform shapes, tube widened from base of corolla to throat if funnelform, but again constricted distally if ventricose or urceolate, outside red, orange, or bright yellow if urceolate, and mostly whitish to pale yellow if funnelform or ventricose but occasionally red, outside mostly glabrous or with an inconspicuous appressed indumentum, occasionally conspicuously pilose, villous, or arachnoid, inside usually glabrous or with an inconspicuous puberulent indumentum, sometimes glandular, color a paler shade of the outside color, if urceolate usually without conspicuous markings, if ventricose usually with yellow, red, purple, or brown spots or lines; spur usually present but occasionally very reduced or absent (but then with a dorsally gibbous tube), apex blunt or pointed, directed straight backward or curved downward, occasionally curved upward; limb mostly the same color as tube but often somewhat darker, lobes mostly subequal, or unequal particularly if corollas are funnelform with a large limb or urceolate with a pouch protruding beyond the limb, mostly rotund, usually small if corolla is urceolate and large if corolla is funnelform, mostly glabrous. Filaments mostly adnate to base of corolla tube, but otherwise free, usually curved or sometimes straight, mostly glabrous, sometimes puberulent; anthers coherent, glabrous, never exserted; one staminode occasionally present. Nectary mostly dorsal and bilobed, less commonly semiannular or annular and thickened dorsally, glabrous to pilose or wooly, frequently directed down into spur. Ovary glabrous or puberulent-velutinous, pale whitish or yellowish; stigma mostly stomatomorphic but often approaching weakly bilobed, sometimes conspicuously bilobed. Fruit a fleshy or semi-fleshy compressed capsule, dehiscing loculicidally with two valves reflexing and disclosing the seeds, sometimes the capsule also splitting septicidally resulting in four lobes at maturity; seeds numerous, mostly elliptic to more or less irregular, irregularly striate and often with papillate projections, usually brown or black.

KEY TO THE SPECIES AND SUBSPECIES OF GASTERANTHUS

1. Corollas hidden by surrounding calyx or often nearly so, funnelform, ventricose, or urceolate (but throat of corolla never surpassed by the inflated, protruding pouch), 1.5–2.5 cm long with 1–3 mm long limb lobes; inflorescences and flowers always glabrous except for occasional minute glandular trichomes on the inside of the calyx/ corolla; stems glabrous (also distally) and often with irregularly wrinkled cortex.
    2. Stems with a fairly smooth cortex; corolla urceolate with a mostly blackish limb; peduncles mostly characteristically wine-red and almost perpendicular to the stem. 15. *G. glaber.*
    2. Stems with an irregularly wrinkled cortex; corolla subventricose to funnelform with a mostly orange limb; peduncles greenish and directed more or less vertically upward rather than being almost perpendicular with the stem.
        3. Fully developed corollas well more than half hidden by the surrounding calyx; spur directed upward; throat with tiny glandular trichomes, 7–9 mm in diameter; nectary villous distally, glabrescent basally. 18. *G. imbricans.*
        3. Fully developed corollas about half hidden by the surrounding calyx; spur bent slightly downward; throat glabrous, 3–6 mm in diameter; nectary glabrous. 7. *G. carinatus.*
1. Corollas never hidden by surrounding calyx or rarely nearly hidden, funnelform, ventricose, occasionally campanulate, or urceolate often with inflated pouch, 0.9–7.5 cm long with limb lobes 1–15 mm long; inflorescences and/or flowers always with an indumentum (in addition to occasional minute trichomes inside the corollas); stems rarely glabrous (at least distally), rarely with a conspicuous irregularly wrinkled cortex.
    4. Corolla ventricose with a relatively small limb and a somewhat constricted throat but never with a ventral pouch near or below the throat, 2.5–4.5 cm long, spur 5–12 mm long, pointed downward, blunt.
        5. Stems and inflorescences with a conspicuous pilose to villous indumentum; leaves mostly with oblanceolate blades; calyx pilose but without glandular trichomes. 34. *G. villosus.*
        5. Stems and inflorescences with an inconspicuous appressed indumentum; leaves mostly with ovate-elliptic blades; calyx with glandular trichomes. 1. *G. acropodus.*
    4. Corolla funnelform to campanulate, or urceolate with a conspicuous ventral pouch near or below the strongly constricted throat, occasionally ventricose but corolla then 2.5 cm or less long and usually with a ventral pouch near the throat, spur 0–30 mm long, direction and apex variable.
        6. Corolla funnelform, occasionally campanulate, and never with a ventral pouch or with a dense, pilose-villous, glandular indumentum outside, usually white to pale yellow and often with darker dots/lines inside, rarely bright yellow or red.
            7. Corolla 1.3–2.5 cm long, red or white to yellow with darker dots, limb lobes 1.5–4 mm long; calyx lobes often with dentate to serrate margin; stems never with villous indumentum.
                8. Corolla yellow-orange-red to red; calyx lobes subentire; peduncles directed nearly vertically upward. 8. *G. columbianus.*
                8. Corolla white to yellow with red or purple spots; calyx lobes with toothed margin; peduncles almost perpendicular to the stem. 35. *G. wendlandianus.*
            7. Corolla 2.2 –7.5 cm long, usually white to pale yellow, rarely with brighter colors (except bright yellow in *G. adenocalyx*, red or bright orange-yellow in *G. mutabilis*, or with dark markings inside in *G. leopardus*), limb lobes 3–15 mm long; calyx lobes variously toothed or entire but rarely with a dentate margin; stem occasionally with villous indumentum.

9. Corolla tube strongly bent downward ventrally 5–10 mm above the attachment of the calyx, throat 15–25 mm in diameter; spur 6–12 mm long, directed strongly downward, 5–10 mm wide nearly to the blunt apex.
10. Inflorescences with peduncles (5–10 mm long); upper leaf surfaces somewhat rugose; nectary glabrous. 12. *G. dressleri.*
10. Inflorescences epedunculate; upper leaf surfaces not rugose; nectary sparsely villous. 16. *G. herbaceus.*
9. Corolla tube not bent strongly downward above attachment of the calyx, throat 6–38 mm in diameter, *or* corolla tube appearing bent and throat 6–9 mm in diameter (*G. epedunculatus*); spur absent or 2–30 mm long, straight, or directed downward and gradually tapering towards pointed apex.
11. Corolla spur 15–30 mm long, strongly downward directed, gradually tapering towards pointed apex; throat 10–38 mm in diameter (upper leaf surface rugose if less than 18 mm); upper leaf surface flat or strongly rugose.
12. Upper leaf surface strongly rugose, blades nearly black above; throat of corolla 12–17 mm in diameter. 4. *G. atratus.*
12. Upper leaf surface flat, blades green above (may be reddish below); throat of corolla 18–38 mm in diameter.
13. Inflorescences pedunculate and mostly 1–3 (–12) -flowered; stems with indumentum at least distally; leaves up to 27 cm long. 11. *G. delphinioides.*
13. Inflorescences epedunculate and 1-flowered; stems glabrous; leaves up to 13 cm long. 30. *G. tenellus.*
11. Corolla spur absent or 2–12 mm long, straight or bent downward, blunt or pointed; throat 6–15 mm in diameter; upper leaf surface flat, or somewhat (but not strongly) rugose.
14. Inflorescences epedunculate but several flowers may develop from leaf axils; corollas without a spur; all calyx lobes directed forward; nectary glabrous. 13. *G. epedunculatus.*
14. Inflorescences pedunculate but occasionally only with a single developed flower; corolla with 2–8 mm long spur; dorsal calyx lobes directed backward surrounding or below spur; nectary appressed pilose to villous.
15. Stomatal clusters conspicuous on lower leaf surfaces, blades with 5–9 secondary veins per side; calyx externally pilose to villous, margin subentire.
16. Calyx lobes ovate to elliptic with acute apex; spur 2–3 mm long. 22. *G. mutabilis.*
16. Calyx lobes narrowly lanceolate with acuminate apex; spur 6–8 mm long. 23. *G. orientandinus.*
15. Stomatal clusters on the lower leaf surfaces inconspicuous, blades with 9–13 (–20) secondary veins per side; calyx externally pilose or conspicuously glandular-villous, margin dentate, serrate, or toothed.
17. Stems glabrous; corolla with downward directed spur; calyx conspicuously glandular-villous particularly outside, margin of lobes irregularly dentate to serrate. 2. *G. adenocalyx.*
17. Stems tomentose-villous, at least distally; corolla with straight or slightly bent spur; calyx pilose to glabrous or glandular outside, margin with more or less conspicuous teeth. 20. *G. leopardus.*
6. Corolla usually urceolate, occasionally ventricose but nearly always with ventral pouch, or when pouch nearly absent, then with a dense pilose-villous glandular indumentum outside, usually orange to red (yellow in *G. osaensis*), and mostly without conspicuous markings inside.
18. Spur directed upward, corolla 3–3.5 cm long; calyx and corolla without conspicuous indumentum but usually with minute glandular trichomes; inflorescences relatively open with up to 4 cm long pedicels. 29. *G. recurvatus.*
18. Spur straight or directed downward, or nearly absent, corolla 0.9–4.5 cm long; calyx and/or corolla mostly with conspicuous indumentum (or glabrous to sparsely puberulent in *G. bilsaensis*, and seldom glabrous in *G. calcaratus*); inflorescences open, or congested with pedicels usually less than 1 cm long.

19. Corolla ventricose to urceolate, throat position distal on tube not being surpassed by protruding pouch, corolla outside often conspicuously pilose or villous, otherwise with sparse or appressed indumentum or glabrous; leaves opposite, blades isophyllous or subisophyllous but never strongly anisophyllous, and never more than 4 times as long as wide.

20. Stems appressed to ground; peduncles nearly absent or up to 2 cm long, shorter than pedicels; corolla 1.8–2.5 cm long and never nearly hidden by the surrounding calyx; filaments basally adnate to corolla tube for 10–12 mm. 3. *G. anomalus.*

20. Stems erect; peduncles up to 20 cm long, usually longer than pedicels; corolla 0.9–3.5 cm long and sometimes nearly hidden by the surrounding calyx; filaments basally adnate to corolla tube up to 6 mm.

21. Corolla nearly hidden by surrounding calyx; calyx lobes lanceolate, red; peduncles not much longer than pedicels. 25. *G. otongensis.*

21. Corolla never hidden by surrounding calyx; calyx lobes ovate, deltate, or rarely lanceolate (in *G. osaensis*), green or rarely orange (in *G. timidus*); peduncles much longer than pedicels.

22. Corolla with appressed indumentum, sparsely puberulent, or glabrous outside, 0.9–2.2 cm long, throat 1.5–4 mm wide; inflorescences densely congested with pedicels rarely exceeding 1 cm.

23. Leaves rugose; stems distally densely villous, trichomes more than 2 mm long. 5. *G. bilsaensis.*

23. Leaves flat, not rugose; stems distally mostly arachnoid-tomentose or villous, trichomes less than 2 mm long. 9. *G. corallinus.*

22. Corolla often conspicuously pilose or villous outside, 1.1–3.5 cm long, throat 3–8 mm wide; inflorescences usually congested with pedicels to 5 mm long (in *G. timidus*) or relatively open with pedicels up to 3 cm long.

24. Inflorescences congested, pedicels up to 5 mm long; corollas 1.1–1.7 mm long, villous outside. 32. *G. timidus.*

24. Inflorescences relatively open, pedicels up to 3 cm long; corolla 2.0–2.5 cm long, densely villous or glandular-pilose outside.

25. Peduncles 4–6 cm long, glabrous to sparsely pilose; calyx lobes entire; corolla yellow; nectary densely villous to glabrescent; ovary puberulent. 24. *G. osaensis.*

25. Peduncles 10–20 cm long, strigillose; calyx lobes serrate; corolla vermilion; nectary glabrous; ovary glabrous. 17. *G. imbaburensis.*

19. Corolla urceolate, throat surpassed by inflated pouch which protrudes beyond the throat (distance from base of corolla to tip of pouch longer than distance from base to throat), corolla outside with indumentum appressed or not, or glabrous but occasionally with a conspicuous pilose to villous to arachnoid indumentum (but then in combination with strongly anisophyllous leaves); leaves opposite or ternate, blades mostly isophyllous but occasionally strongly anisophyllous, occasionally more than 4 times as long as wide.

26. Leaf blades with 13–18 secondary veins per side, maximum distance between secondary veins 7 mm but usually less than 5 mm, upper surface usually somewhat rugose between the parallel secondary veins and indumentum restricted to rugose bands between the veins; corollas 1.2–2.5 cm long. 28. *G. quitensis.*

26. Leaf blades usually with 5–11 secondary veins per side but occasionally up to 14; maximal distance between secondary veins often exceeding 7 mm, indumentum on upper leaf surfaces not located in rugose bands between parallel veins; corollas 1.2–4.5 cm long.

27. Calyx mostly glabrous or nearly glabrous, at most sparsely strigillose (in *G. trifoliatus*), often (or mostly) with conspicuous lighter green dots or blotches particularly distally, occasionally arachnoid or appressed-pilose and/or without dots (and shape of lobes then narrowly lanceolate); inflores-

cences with either 1 or 2 (rarely to 4) flowers; leaves opposite or ternate, occasionally up to 8 times as long as wide.

28. Calyx lobes narrowly lanceolate with long-acuminate apex, indumentum arachnoid, not with lighter green dots or blotches; blades of leaves often more than 3.5 times as long as wide. 27. *G. perennis.*

28. Calyx lobes ovate, occasionally lanceolate, apex acute, occasionally acuminate, indumentum absent or inconspicuous, lighter green dots or blotches sometimes present; blades of leaves maximally 3.5 times as long as wide (or up to 8 times longer in *G. trifoliatus*).

29. Leaves opposite, blades usually more than 4 cm wide; inflorescences always 1-flowered; corolla yellow-orange or reddish-orange. 26. *G. pansamalanus.*

29. Leaves ternate (rarely 2 or 4 at a node), blades usually less than 4 cm wide; inflorescences with 1 or 2 flowers; corolla red.

30. Blades elliptic, obovate, or oblanceolate, 3–13 cm long, 1–5 cm wide, petioles 0.5–2.5 cm long; calyx lobes glabrous, ovate to elliptic; nectary bilobed. 31. *G. ternatus.*

30. Blades lanceolate to oblanceolate, 10–12 cm long, 1.5–2.0 cm wide, petioles 0.5–1.0 cm long; calyx lobes sparsely strigillose, ovate to broadly ovate; nectary semiannular. 33. *G. trifoliatus.*

27. Calyx lobes mostly with a conspicuous indumentum but without conspicuous lighter green dots; inflorescences with 2 or more flowers (but sometimes only 1 or 2 develop); leaves opposite, rarely ternate (but then the inflorescences with more than 2 flowers), maximally 3.5 times as long as wide (except occasionally when leaves also are strongly anisophyllous).

31. Stomatal clusters on lower leaf surfaces large and conspicuous, between clusters a pilose or villous indumentum, blades maximally 14 cm long and either obovate-orbicular or elliptic; inflorescences with 1–4 flowers; corollas 2.8–4.0 cm long.

32. Leaves obovate to orbiculate, stems glabrous to more or less tomentose distally. 10. *G. crispus.*

32. Leaves elliptic; stems conspicuously pilose to villous at least distally. 14. *G. extinctus.*

31. Stomatal clusters on lower leaf surface conspicuous or not to the naked eye (*G. calcaratus*) but sometimes not very distinct, lower surface indumentum mostly appressed and restricted to veins; blades up to 26 cm long, mostly ovate or oblanceolate, occasionally elliptic, obovate, or falcate; inflorescences with 2–10 (–25) flowers; corollas 1.2–4.5 cm long.

33. Corollas 2–4.5 cm long; leaves with obovate to oblanceolate blades, rarely elliptic, with more or less attenuate bases; petioles mostly less than 1.5 cm long; inflorescences with a conspicuous villous, arachnoid, or woolly indumentum.

34. Calyx lobes 12–22 mm long, lanceolate with acuminate apex; upper part of peduncle blackish and glabrous contrasting with lower part having arachnoid-woolly indumentum. 21. *G. macrocalyx.*

34. Calyx lobes 4–12 mm long, ovate with acute to obtuse apex; inflorescences with uniform indumentum. 19. *G. lateralis.*

33. Corollas 1.2–2.5 cm long but occasionally to 3.5 cm (in Colombia); leaves ovate to elliptic, occasionally oblanceolate, elliptic, or falcate, usually with cuneate to acute bases; petioles 0.5–3 cm long, but mostly exceeding 1.5 cm; inflorescences mostly without conspicuous indumentum. 6. *G. calcaratus.*

35. Leaves mostly strongly anisophyllous but sometimes subequal, blade often falcate and up to 6 times as long as wide; flowers often with conspicuous villous indumentum on calyces and corollas; nectary glabrous. 6b. *G. calcaratus* subsp. *calceolus.*

35. Leaves usually isophyllous, blades never falcate and maximally 3.5 times as long as wide; flowers without conspicuous villous indumentum; nectary pubescent to villous.
36. Leaves usually with 9–12 secondary veins per side, blades mostly elliptic; corollas occasionally up to 3.5 cm long. 6a. *G. calcaratus* subsp. *calcaratus.*
36. Leaves usually with 5–8 secondary veins per side, blades mostly elliptic-oblanceolate; corolla length not exceeding 2.5 cm. 6c. *G. calcaratus* subsp. *oncogastrus.*

1. Gasteranthus acropodus (J. D. Smith) Wiehler, Selbyana 1: 155. 1975. *Besleria acropoda* J. D. Smith, Bot. Gaz. 54: 240. 1912.—TYPE: COSTA RICA. Limón: Tsaki, Talamanca, 200 m, Apr 1895, *Tonduz 9554* (holotype: US!; isotype: CR!).

Besleria allenii C.V. Morton, Ann. Missouri Bot. Gard. 29: 38. 1942. *Gasteranthus allenii* (C. V. Morton) Wiehler, Selbyana 1: 155. 1975.—TYPE: PANAMA. Coclé: Vicinity of La Mesa, 1000 m, 12 May 1941, *Allen 2371* (holotype: US!; isotypes: A! EAP! MO! PMA! S!).

Besleria allenii var. *paucivenia* C. V. Morton, Ann. Missouri Bot. Gard. 29: 38. 1942.—TYPE: PANAMA. Coclé: Hills N of El Valle de Antón, 1000 m, 14 Jul 1940, *Allen 2188* (holotype: US!; isotype: PMA!).

Subshrubs to shrubs; stems erect, usually to 1 m, rarely to 5 m tall, to 6 mm in diameter, glabrous to inconspicuously strigillose distally, internodes 1–4 cm long, terete to quadrangular. Leaves opposite, subisophyllous; blades 7–20 cm long, 3–7 cm wide, ovate to elliptic or obovate, base attenuate to acute or obtuse and decurrent, apex acute to acuminate, margin serrate, crenate, or subentire, flat, adaxially dark green, glabrous, abaxially pale green, more or less strigillose and mainly on the veins, lateral veins 7–8 per side, stomatal clusters often conspicuous; petioles 1–3 cm long, strigillose to glabrous. Inflorescences open cymes of 1–5 flowers, inconspicuously and sparsely pilose; peduncles 2–6 cm long; pedicels 0.5–2.0 cm long. Calyx whitish, pale green, or purplish, glandular-pilose to glandular-villous particularly outside, lobes nearly free to base, subequal to unequal, each of the dorsal and lateral lobes ovate to lanceolate, ventral lobes more lanceolate, apex mostly acute, sometimes acuminate, dorsal lobe keeled, 6–15 mm long, 3–6 mm wide, serrate to nearly entire, dorsal lobe directed backward below spur, ventral lobes directed forward and appressed to corolla, lateral lobes directed upward. Corolla 2.5–4.5 cm long, ventricose with a relatively small limb, tube 1.5–2.6 cm long, widened from 3–5 mm at base to 1.1–1.7 cm in diameter at widest point, mainly yellow, but varying from cream-white to lemon-yellow, outside more or less appressed-pilose, inside puberulent distally, markings with or without reddish lines, spots, or patches; spur 6–12 mm long, to 5 mm wide, apex blunt, directed downward; throat 9–13 mm in diameter, puberulent; limb yellow, lobes 2–7 mm long, 5–7 mm wide, subequal-unequal, acute-rotund, glabrous. Filaments 10–14 cm long, adnate to base of corolla tube for 2–4 mm, glabrous; anthers ca. 1.5 mm long, 1 mm wide. Nectary dorsal, bilobed, to 2 mm high, and directed into spur, tomentose distally, glabrescent basally. Ovary ca. 4 mm long, 2.5 mm in diameter, glabrous to sparsely puberulent; style ca. 7 mm long, puberulent or glabrous; stigma stomatomorphic to weakly bilobed. Capsule 5 mm long, 8 mm wide, oblate-subglobose, bivalved; seeds ca. 0.3 mm long, 0.2 mm wide, oblong, obliquely striate. Figs. 9C, 13E, F.

Phenology. Collected in flower during every month of the year; in fruit in January.

Distribution (Fig. 15). Eastern Costa Rica (Limón) and Panama (Bocas del Toro,

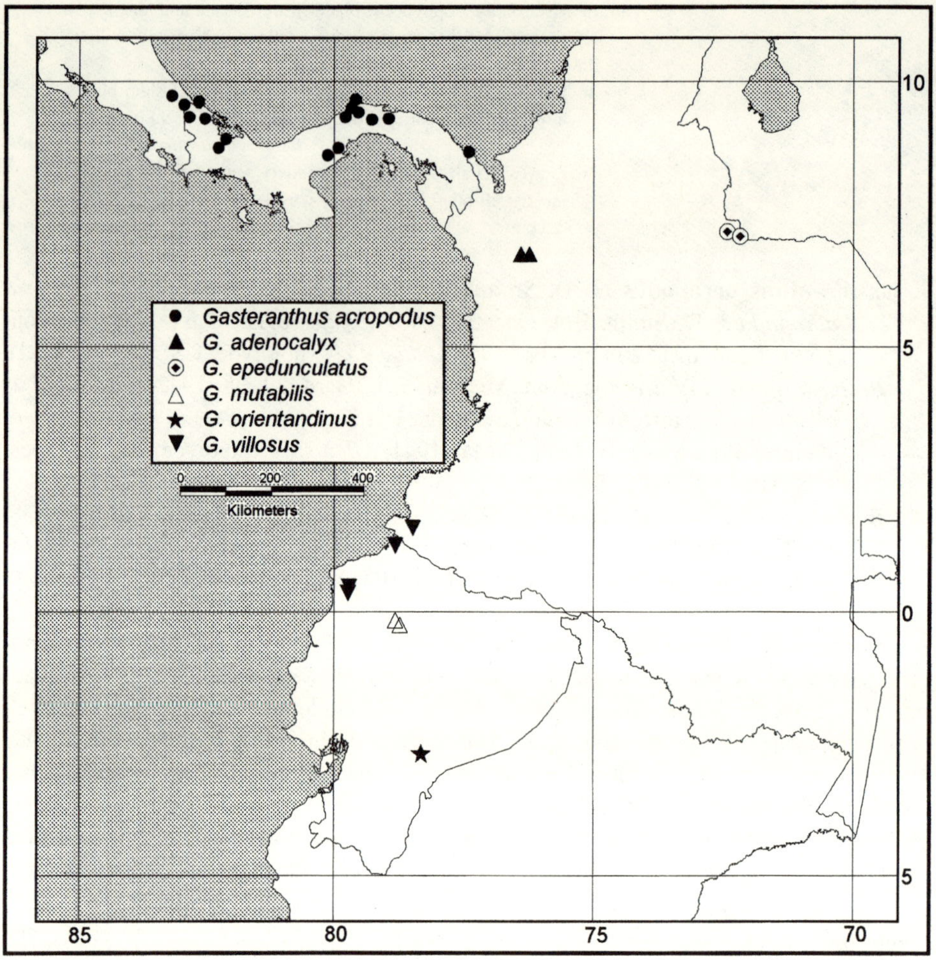

FIG. 15. Distribution of *Gasteranthus acropodus*, *G. adenocalyx*, *G. epedunculatus*, *G. mutabilis*, *G. orientandinus*, and *G. villosus*.

Chiriquí, Coclé, Colón, Panamá, and San Blas); lowland wet forest to cloud forests, usually near streams or in other wet habitats; 0–1000 m.

REPRESENTATIVE SPECIMENS. **Costa Rica.** LIMÓN: large swamp at Manzanillo de Talamanca, *Grayum & Burton 4317* (MO); Tortuguero Cantón, Suretka, *Barringer et al. 3608* (F). **Panama.** BOCAS DEL TORO: between Fortuna and Chiriquí Grande, 2.2 mi N of continental divide, 6.3 mi N of bridge over Fortuna Lake, *Croat 60403* (MO); Milla 7.5, ridge SW of railroad station, *Kennedy & Dressler 1275* (SEL, US).—CHIRIQUÍ: Fortuna Dam area to Chiriquí Grande, *D'Arcy et al. 15989* (MO, US).—COCLÉ: Cerro Gaital Caracoral nr. Cerro Pilón, *Dwyer & Correa 8851* (US); El Valle, *Wiehler 71270* (SEL, US).—COLÓN: along banks of Río Guanche, *Sullivan 146* (NY, US); Río Iguanita, ca. 2 mi from the sea, *D'Arcy et al. 13341* (US).—PANAMÁ: Cerro Campana, along trail to summit, *Croat 17231* (SEL, US-2 sheets).—SAN BLAS: Nusagandi, *McDonagh et al. 669* (MO).

Gasteranthus acropodus has medium-sized (mostly 3–4 cm long) ventricose corollas with a relatively small limb and a blunt downward bent spur (Fig. 9C). *Gasteranthus villosus* from Ecuador has similar corollas (Fig. 9D) but differs by being much more pilose to villous (see *G. villosus*, no. 34); *G. osaensis* from the Osa peninsula of southeastern Costa Rica is vegetatively similar. The latter species, however, has smaller and more urceolate corollas (Fig. 10A; also see *G. osaensis*, no. 24).

Gasteranthus acropodus is primarily a species of the Caribbean slope in Costa Rica and Panama. Only two collections are known from the Pacific slope, one from the province of Chiriquí near the border with Bocas del Toro (*D'Arcy et al. 15989*), and the other from Cerro Campana in the province of Panamá (*Croat 17231*).

2. Gasteranthus adenocalyx L. E. Skog & L. P. Kvist, sp. nov.—TYPE: COLOMBIA. Antioquia: Mpio. Frontino, Corregimiento La Blanquita, Murrí region, Alto de Cuevas, 14.5 km from Nutibara along road to La Blanquita, 1850 m, 14 Jul 1988, *Callejas et al. 6800* (holotype: HUA!; isotype: MO!).

A Gasterantho epedunculato L. E. Skog & L. P. Kvist, G. mutabili L. E. Skog & L. P. Kvist, et G. orientandino L. E. Skog & L. P. Kvist lobis calycis dentatis vel serratis villosis et glanduliferis differt.

Herbs; stems erect, to 1 m tall, to 6 mm in diameter, glabrous, internodes 1.5–3 cm long, quadrangular, somewhat longitudinally wrinkled. Leaves opposite, isophyllous; blades 9–16 cm long, 5–7 cm wide, oblique, ovate to elliptic, base acute and occasionally oblique, apex acute, margin crenate to subentire, somewhat rugose, minutely muricate, adaxially dark, dull green with considerable lighter primary and secondary veins, sparsely appressed-pilose to glabrous, abaxially green, with conspicuous darker veins, glabrous but minutely papillate, lateral veins 11–12 per side, stomatal clusters not conspicuous; petioles 2–3 cm long, glabrous. Inflorescences open cymes of 3–12 flowers, glabrous; peduncles 5–8 cm long; pedicels 0.5–3 cm long. Calyx green, glandular-villous particularly outside, lobes basally connate 1–3 mm, unequal, each of the lobes narrowly ovate to broadly ovate, apex acute, dorsal lobe keeled, dorsal lobe ca. 8 mm long, 5 mm wide, ventral lobes ca. 5 mm long, 5 mm wide, lateral lobes ca. 8 mm long, 8 mm wide, dentate to serrate, dorsal lobe directed backward surrounding spur, ventral lobes directed forward and appressed to corolla, lateral lobes directed upward. Corolla 3–3.5 cm long, funnelform with large limb, tube not bent downward, 1–1.5 cm long, widened from 3–4 mm at base to 9–13 mm in diameter at widest point, bright yellow, outside dorsally pilose to villous, sometimes glandular, inside glandular-puberulent in throat near base of tube, markings with ca. 5 red stripes from throat; spur 4–8 mm long, to 3 mm wide, pointed, directed downward; throat 9–13 mm in diameter; limb yellow, lobes 5–7 mm long, 6–9 mm wide, subequal, rotund, glabrous. Filaments 8–13 mm long, adnate to base of corolla tube for 2–4 mm, sparsely puberulent basally; anthers ca. 2 mm long, 1.5 mm wide. Nectary dorsal, bilobed, ca. 2 mm high and directed down into spur, distally villous, tomentose basally. Ovary 3–4 mm long, 2–3 mm in diameter, puberulent; style 8–10 mm long, puberulent; stigma stomatomorphic. Capsule ca. 3 mm long, ca. 8 mm wide, flattened, pattern of dehiscence unknown; seeds ca. 0.3 mm long, 0.2 mm wide, elliptic, papillate. Figs. 2B, 16.

Phenology. Collected in flower in January, March, April, and June; in fruit only during March.

Distribution (Fig. 15). Colombia (Antioquia near the border with Chocó); stream banks in wet forest; 1700–2000 m.

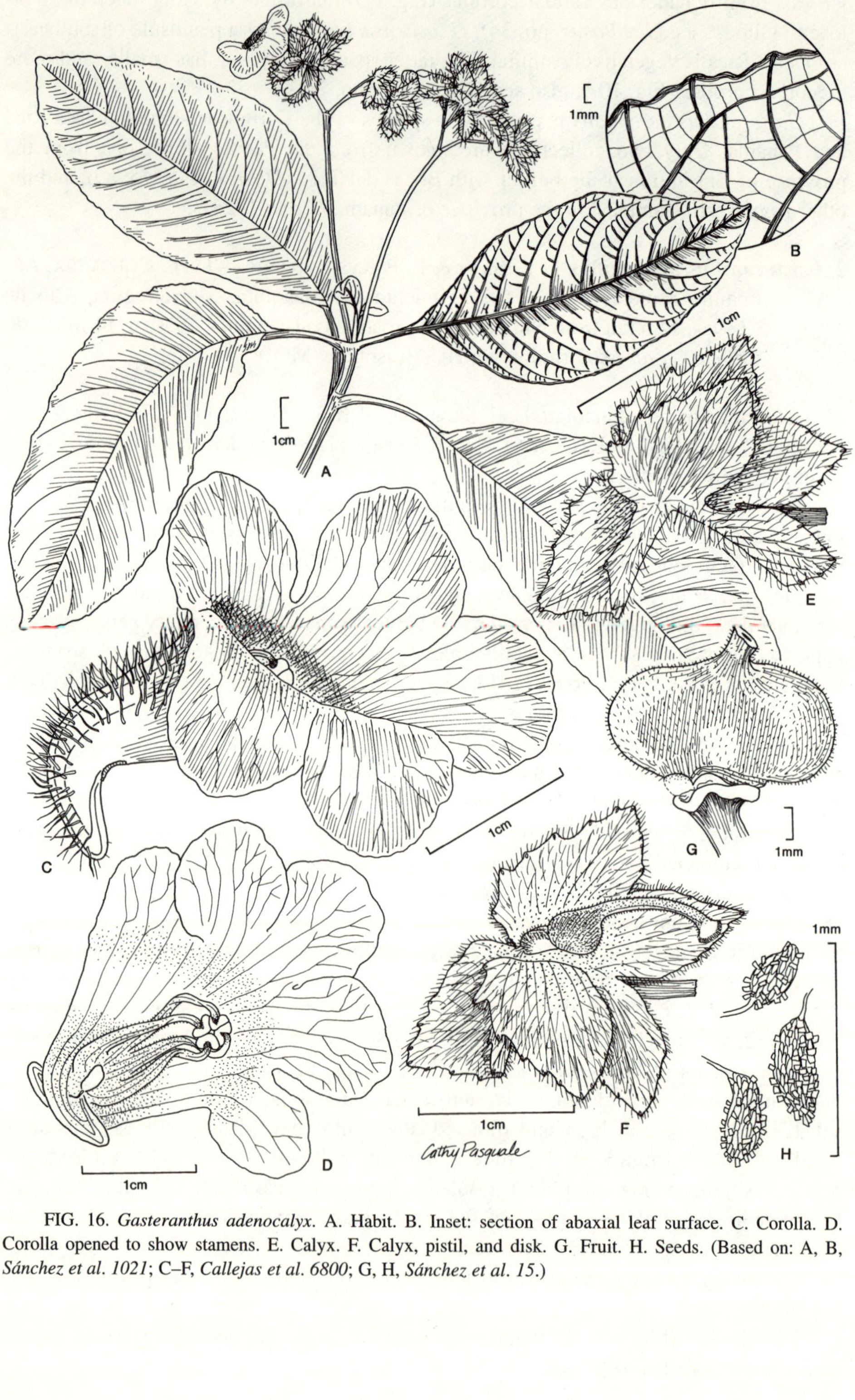

FIG. 16. *Gasteranthus adenocalyx*. A. Habit. B. Inset: section of abaxial leaf surface. C. Corolla. D. Corolla opened to show stamens. E. Calyx. F. Calyx, pistil, and disk. G. Fruit. H. Seeds. (Based on: A, B, *Sánchez et al. 1021*; C–F, *Callejas et al. 6800*; G, H, *Sánchez et al. 15*.)

ADDITIONAL SPECIMENS EXAMINED. **Colombia.** ANTIOQUIA: Mpio. Frontino, Correg. Nutibara, Murrí region, rd towards La Blanquita, *Acevedo et al. 1333* (HUA); Correg. Nutibara, Vereda Alta de Cuevas, Quebrada del Oso, *Fonnegra 3513* (HUA); Correg. Nutibara, Río Cuevas valley, *Sánchez et al. 15* (MEDEL), *1021* (US).

Gasteranthus adenocalyx has medium-sized funnelform corollas (3–3.5 cm long) with a wide limb (Fig. 2B). It is characterized by the contrast between the densely glandular-hairy calyx lobes and the large bright yellow, glabrous limb, and the equally glabrous pedicels and peduncles (Fig. 16). Other distinctive characteristics include the rather irregularly dentate margin of the calyx lobes, the conspicuous lighter primary and secondary veins on the upper leaf surfaces, as well as the muricate upper leaf surface. The species most similar to *G. adenocalyx* are *G. epedunculatus* (no. 13), *G. mutabilis* (no. 22), and *G. orientandinus* (no. 23), but they all have calyx lobes with entire or subentire margins and without glandular hairs (Fig. 2).

3. Gasteranthus anomalus (C. V. Morton) Wiehler, Selbyana 1: 154. 1975. *Besleria anomala* C. V. Morton, Contr. U.S. Natl. Herb. 26: 471. 1939.—TYPE: COLOMBIA. Boyacá (?): Quebrada de la Honda, Mar 1846, *Purdie s.n.* (holotype: K!; isotypes: K-3 sheets!, fragment: US!).

Herbs; stems appressed to ground, usually very short and bearing leaves nearly in a rosette, but occasionally to 20 cm tall, to 6 mm in diameter, pilose to villous, internodes to 5 mm long, rather irregular in cross section. Leaves opposite, often congested, isophyllous; blades 6–21 cm long, 5–12 cm wide, obovate, base obtuse to weakly truncate, apex obtuse, margin crenate to serrate, flat, adaxially dark green, glabrous but weakly muricate, abaxially lighter green, pilose to villous on the veins, lateral veins 5–7 per side, stomatal clusters not conspicuous; petioles 1–4 cm long, pilose to villous. Inflorescences open cymes of 1–6 flowers, pilose, rarely villous; peduncles much reduced, 0.2–2 cm long; pedicels (with developed flowers) 2–4 cm long. Calyx green, mostly sericeous, lobes basally connate for 0.5–2 mm, equal, each of the lobes 7–12 mm long, 2–4 mm wide, lanceolate with acute apex, subentire, all lobes directed forward. Corolla 1.8–2.5 cm long, urceolate with a large limb and a ventral pouch not extending beyond throat, tube 1.4–2 cm long, widened from 2–4 mm at base to 0.7–1.1 cm (with pouch) in diameter at widest point, red, outside appressed-pilose, particularly dorsally, inside puberulent in upper part, markings not evident; spur absent (but tube dorsally gibbous); throat 4–7 mm in diameter; limb red, lobes 2–3 mm long, 2–3 mm wide, subequal, rotund, puberulent to glabrous along edge. Filaments 13–18 mm long, adnate to base of corolla tube for 10–12 mm, pilose; anthers ca. 1.5 mm long, 2.5 mm wide. Nectary thickened dorsally, bilobed to annular, to ca. 1 mm high, villous distally, glabrescent basally. Ovary ca. 4.5 mm long, 2.5 mm in diameter, velutinous; style ca. 13 mm long, puberulent; stigma stomatomorphic. Capsule ca. 6 mm long, ca. 6 mm wide, flattened, deshiscence pattern unknown; seeds not seen. Figs. 10B, 11B.

Phenology. Collected in flower from November to January; in fruit during December.

Distribution (Fig. 19). Colombia (Antioquia, Boyacá?); riverbanks and wet places; 300–800 m.

ADDITIONAL SPECIMENS EXAMINED. **Colombia.** ANTIOQUIA: Mpio. San Luís, Cañón del Río Claro, *Cogollo 980* (COL, MO), *1279* (MO); Mpio. San Luís, Quebrada "La Cristalina," *Ramírez & López 324* (HUA, MO); Río Claro, *Rentería et al. 4955* (HUA).—LOCALITY UNKNOWN: "N. Granada," *Purdie s.n.* (GH).

Gasteranthus anomalus is easily set apart from all other species in the genus by the combination of shoots with congested leaves appressed to the ground, red urceolate corollas, and inflorescences with peduncles nearly absent to much shorter than the pedicels. Several species with large, whitish funnelform corollas may also have shoots appressed to the ground (see *G. atratus*, no. 4; *G. dressleri*, no. 12; and *G. herbaceus*, no. 16), but the only other species in the genus with red, urceolate corollas and stems that may also be decumbent is *G. crispus* (no. 10). That species, however, differs from *G. anomalus* by having peduncles longer than the pedicels, larger flowers, and very conspicuous stomatal clusters on the lower leaf surfaces. An additional distinctive character is the lack of a spur in the corollas of *G. anomalus* (Fig. 10B). This character is shared only with *G. epedunculatus*, which has funnelform, whitish corollas (Fig. 2A).

Apparently *Gasteranthus anomalus* is endemic only to a small area in the Río Magdalena valley of Colombia. The few recent collections come from the same canyon along the Río Claro in the department of Antioquia. The type collection made by Purdie may have come from the department of Boyacá.

4. Gasteranthus atratus Wiehler, Selbyana 5: 86, pl. 3C. 1978.—TYPE: Plants cultivated from living material collected by Madison in 1977 and also by Dodson in 1977, originally from Montañas de Ila, cloud forest along ridge line near La Centinela, Km 12, on road from Patricia Pilar to Flor de Mayo, Provinces of Pichincha or Los Ríos, Ecuador, *Wiehler 78166* (holotype: SEL, not located); same locality, 16 Jul–11 Aug 1977, *Dodson & Dodson 6838* (lectotype, here designated: SEL!).

Herbs; stems erect, to 50 cm tall, to 5 mm in diameter, glabrous to somewhat papillose, internodes 1–3 cm long, mostly terete, occasionally quadrangular. Leaves opposite, subisophyllous; blades 6–14 cm long, 3–7 cm wide, ovate, base acute to obtuse, apex acuminate, margin irregularly dentate, rugose with up to 5 mm high bullate ridges along veins, adaxially shiny blackish green, glabrous to somewhat papillate, abaxially purplish, pilose to villous, particularly on the veins, lateral veins 8–9 per side, stomatal clusters conspicuous; petioles 1–5 cm long, glabrous to sparsely pilose. Inflorescences open cymes of 2–4 flowers, glabrous; peduncles 3–6 cm long; pedicels 1–2 cm long. Calyx green, outside papillate basally, distally glabrous, lobes nearly free to base, subequal, each of the lobes ovate, with acute apex; dorsal lobe keeled, 12–17 mm long, 8–12 mm wide, crenate to dentate, dorsal lobe directed backward and down below spur, ventral lobes directed forward and appressed laterally to corolla, lateral lobes directed upward. Corolla 4.5–6.0 cm long, funnelform with a large limb, tube 2.8–3.5 cm long, widened from 3–4 mm at base to 12–17 mm in diameter at widest point, cream to yellow, outside glabrous, inside glabrous, with red stripes; spur 15–20 mm long, 1–3 mm wide, pointed, bent downward; throat 12–17 mm in diameter; limb pale yellow, lobes 3–5 mm long, 7–10 mm wide, subequal, rotund, glabrous. Filaments 11–17 mm long, adnate to base of corolla tube for 3–5 mm, glabrous; anthers ca. 1.5 mm long, 1.5 mm wide. Nectary dorsal, bilobed, to 2 mm high, pilose. Ovary ca. 5 mm long, 5 mm in diameter, glabrous to sparsely puberulent; style ca. 15 mm long, glabrous; stigma stomatomorphic. Capsule ca. 1.2 cm long, 1 cm wide, shape unknown, bivalved, partly splitting secondarily into 4 valves; seeds 0.5 mm long, 0.3 mm wide, oblong, covered with spine-like projections. Fig. 3A.

Phenology. Collected in flower in February, July, August, and November; in fruit during November.

Distribution (Fig. 24). Ecuador (Cotopaxi, Los Ríos, and Pichincha); undisturbed moss- and epiphyte-enshrouded cloud forest on steep slopes and ridges, in deep shade; 300–1000 m.

ADDITIONAL SPECIMENS EXAMINED. **Ecuador**. COTOPAXI: Hacienda Solento, near Santa Rosa, Cantón Pujilí, *Mexia 6715* (US).—LOS RÍOS OR PICHINCHA: Centinela, *Dodson et al. 7609* (MO, SEL); Centinela, *Dodson & Dodson 15918* (US); Centinela, *Iltis et al. E114* (SEL, WIS); Centinela, *van der Werff et al. 12394* (MO, QCNE).

The nearly black, deeply undulate, rugose or bullate leaves distinguish *G. atratus* from all species in the genus with similar large and whitish funnelform corollas. Only *G. bilsaensis* (Fig. 17) has similar leaves, but this species has small, red, urceolate corollas. Both species occur in low-elevation cloud forest but they are not sympatric (see *G. bilsaensis*, no. 5). Another similar species is *G. tenellus*, also known from western Ecuador, but the two taxa are apparently not sympatric. *Gasteranthus tenellus* occurs about 100 km to the northwest in the coastal hill range in Esmeraldas province. That species also has a funnelform corolla (Fig. 3C) with a long, pointed, and downward-bent spur, but it never has rugose leaves or decumbent shoots with the leaves appressed to the ground. *Gasteranthus delphinioides*, from Colombia, Panama, and Costa Rica, is also similar but is a more robust plant with larger flowers (Fig. 3B).

Gasteranthus atratus is only known from a small area along the foothills of the Andes in western Ecuador. All collections apparently come from Centinela Ridge near the Los Ríos/Pichincha border, except for *Mexia 6715*, which was collected east of Centinela Ridge on the western Andean slopes in Cotopaxi province.

The holotype of *Gasteranthus atratus* has not been located at SEL (Holst, pers. comm.), nor have any of the isotypes been found. We select the only other specimen cited in the protologue as a lectotype.

5. Gasteranthus bilsaensis L. E. Skog & L. P. Kvist, sp. nov.—TYPE: ECUADOR. Esmeraldas: Montañas de Mache, 35 km W of Quinindé, Bilsa Biological Station, Rana Roja trail, 550–650 m, 25 Oct 1995, *Mendoza-T. et al. 561* (holotype: QCNE!; isotypes: AAU! COL! QCA! SEL! US!).

A Gasterantho corallino (Fritsch) Wiehler statura usque ad 40 cm, foliis minoribus laminis 7–15 cm longis valde rugosis, corollis usque ad 1.4 cm longis, faucibus 1.5–2.5 mm diametro, pedunculis subnulli vel usque ad 6 cm longi, et caulibus petiolis inflorescentiis dense villosis differt.

Herbs; stems erect, 20–40 cm tall, to 5 mm in diameter, glabrescent below to distally densely villous with trichomes 2 mm or more long, internodes 1–2.5 cm long, terete to quadrangular. Leaves opposite, isophyllous, blades 7–15 cm long, 3–7 cm wide, ovate to elliptic or oblanceolate, base cuneate to acute, sometimes oblique, apex acute, margin serrate, rugose with more than 1 mm high bullate ridges along main veins, adaxially dark green to nearly black, pilose, abaxially lighter green, pilose to villous on the veins, lateral veins 8–11 per side, stomatal clusters not conspicuous; petioles 0.5–2.5 cm long, glabrous to arachnoid. Inflorescences congested cymes of 8–20 flowers, pilose to villous; peduncles nearly absent to 6 cm long; pedicels 1–5 mm long. Calyx dark green, outside pilose, especially along veins, to glabrescent; inside velutinous; margins pilose, lobes basally connate ca. 1 mm, subequal, each of the lobes 5–7 mm long, 3–6 mm wide, ovate with acute apex, serrate, dorsal lobe directed backward surrounding spur, ventral lobes directed forward, lateral lobes directed upward. Corolla 0.9–1.4 cm long, urceolate with small limb and ventral pouch not exceeding throat, tube 0.7–1.0 cm, widened from 1–2 mm at base to 3–7 mm (at pouch) in diameter at widest point, orange-red, outside sparsely pubescent to glabrous, inside pilose in throat; an inner layer of tissue can be liberated from outer layer, markings none; spur 2–3 mm long, 2–4 mm wide, blunt, straight; throat 1.5–2.5 mm in diameter; limb red, lobes 1–2 mm long, 2–3 mm wide, equal, rotund, glabrous. Filaments 6–9 mm long, adnate to base of corolla tube for 2–3 mm, glabrous; anthers ca. 2.5 mm long, 2 mm wide. Nectary dorsal, semiannular, to 1 mm high, densely pubescent. Ovary 3–4 mm long, 3–4 mm in diameter, puberulent; style ca. 5 mm long, glabrous; stigma stomatomorphic. Capsule ca. 4 mm long, ca. 6 mm wide, flattened, bivalved; seeds not seen. Figs. 7D, 17.

Phenology. Collected in flower in October and November; in fruit during October.

Distribution (Fig. 22). Ecuador (Esmeraldas, Manabí); cloud forest or wet primary and secondary forest, near streams and waterfalls; 300–700 m.

ADDITIONAL SPECIMENS EXAMINED. **Ecuador**. ESMERALDAS: Mache mountains, 35 km W of Quinindé, Bilsa Biological Station, *Clark et al. 213* (AAU, COL, E, MO, QCNE, US), *Clark et al. 1654* (COL, E, MO, QCNE, US); Bilsa, *Pitman et al. 884* (MO, QCNE, US); Bilsa, *Mendoza-T. et al. 571* (AAU, COL, QCA, QCNE, US).—MANABÍ: 10 km E of Pedernales, Cerro Pato de Pájaro, *Clark et al. 2716* (US).

The combination of rugose leaves and very small, red, urceolate corollas in densely congested inflorescences sets *G. bilsaensis* apart from all other species. It is similar to *G. corallinus*, and may share a common ancestor with this widely distributed species in the isolated cloud forests of the upper Cordillera de Mache. The two species are not sympatric (see Fig. 22), although *G. corallinus* also is found in the province of Esmeraldas, but apparently only in forests north of the Esmeraldas River, e.g., in lowland rainforests in the Cayapa-Onzole river system (pers. obs.). *Gasteranthus bilsaensis* differs primarily from *G. corallinus* by having strongly rugose rather than flat upper leaf surfaces; however, a combination of other characters also sets *G. bilsaensis* apart from *G. corallinus*; viz., a dense villous indumentum composed of trichomes more than 2 mm long on the stems, petioles, and inflorescences; mostly very short peduncles that never become longer than 5 cm; small corollas with a narrow throat (maximally 1.4 cm long and 2.5 mm wide, but usually smaller); and small stature (maximally 40 cm tall) and relatively small leaves (maximally 15 cm long).

Gasteranthus bilsaensis is known from the coastal hill range near the Pacific coast of Ecuador in the provinces of Esmeraldas and Manabí. The species is locally common in the Bilsa Biological Reserve in the Cordillera de Mache, and has also been collected recently farther to the south in Manabí on Cerro Pato de Pájaro.

6. Gasteranthus calcaratus (Kunth) Wiehler, Selbyana 1: 154. 1975. *Besleria calcarata* Kunth in H.B.K., Nov. gen. sp. pl. 2: 399 (qto. ed.), 320 (fol. ed.). 1818. *Alloplectus calcaratus* (Kunth) G. Don, Gen. hist. 4: 655. 1838.—TYPE: COLOMBIA. Quindio (?): La Pamilla, 1130 hex., Oct, *Humboldt & Bonpland 1878 p.p.* (holotype: P-Bonpl!).

Herbs to subshrubs; stems erect, to 1.5 m tall (to 3 m tall?; note on one collection label), to 7 mm in diameter, glabrescent to tomentose or arachnoid distally, sometimes only sparsely pilose to glabrous distally, internodes either terete or quadrangular, 0.5–5

FIG. 17. *Gasteranthus bilsaensis*. A. Habit. B. Inset: section of abaxial leaf surface. C. Corolla. D. Corolla opened to show stamens. E. Calyx. F. Calyx, pistil, and disk. (Based on: A, *Mendoza-T. et al. 561*; B, *Clark et al. 213*; C–F, *Pitman et al. 884*.)

(–9) cm long. Leaves opposite, rarely ternate, mostly isophyllous but in some populations strongly anisophyllous; blades 8–18 (–22) cm long, 2.5–10 cm wide, most commonly ovate to elliptic or oblanceolate but in some populations narrowly lanceolate-falcate, base cuneate to acute, in some populations conspicuously oblique, apex acuminate, sometimes acute, margin mostly remotely serrate, less commonly dentate to subentire, flat, rarely somewhat undulate, adaxially green, glabrous, rarely appressed pilose, abaxially lighter green, pilose to villous or tomentose usually only at the veins, lateral veins 5–14 per side but little variation within populations, lateral veins usually more than 7 mm apart, stomatal clusters conspicuous in most populations but sometimes not very distinct; petioles 0.5–3 cm long, mostly arachnoid-tomentose. Inflorescences mostly congested or sometimes relatively open cymes of 2–10 (–25) flowers, mostly glabrous, occasionally but inconspicuously pilose to villous; peduncles 3–13 cm long; pedicels usually less than 0.8 cm long and maximally 1.5 cm long. Calyx mostly green, sometimes purplish, outside mostly glabrous or sparsely pilose, some populations pilose to villous and/or glandular, lobes basally connate 1–2 mm, subequal to unequal, each of the lobes mostly ovate with acute apex, less commonly suborbicular with rotund apex, or lanceolate with acuminate apex; dorsal lobe usually keeled, 2–6 (–10) mm long, 2–7 mm wide, subentire, often ciliate, rarely somewhat serrate, dorsal lobe directed backward below spur, ventral lobes directed forward, lateral lobes directed upward. Corolla 1.2–2.5 (–3.5) cm long, urceolate with a small limb and a ventral pouch, tube (defined as distance from base of corolla to throat) 0.8–1.8 cm but distance from base to tip of pouch 1–3 cm, width from 1.5–4 mm at base to 4–12 mm (with pouch), orange-red, outside puberulent to glabrous, rarely pilose to villous, inside glabrous, markings not seen; spur 2–5 mm long, 2–4 mm wide, blunt, straight; throat 2–5 mm in diameter; limb orange-red, often darker than outside of corolla, lobes 1–3 mm long, 2–4 mm wide, subequal-unequal, rotund, dorsal lobes the largest, glabrous. Filaments 6–15 mm long, adnate to base of corolla tube for 2–3 mm, mostly glabrous, sometimes puberulent; anthers ca. 2 mm long, 1.5 mm wide. Nectary thickened dorsally, bilobed to semi-annular, ca. 1 mm high, glabrous or pubescent to villous. Ovary 3–4 mm long, 3–4 mm in diameter, glabrous-puberulent; style 6–8 mm long, glabrous; stigma stomatomorphic to bilobed. Capsule 0.4–0.5 mm long, 6–9 mm wide, flattened, bivalved, secondarily splitting into 4 valves; seeds 0.3–0.4 mm long, 0.2 mm wide, ellipsoid or ovoid, papillate. Figs. 6, 12B–D, G–J.

Gasteranthus calcaratus is the most variable of all species in the genus. The following combination of characters applies to the vast majority of specimens of *G. calcaratus*: 1) corollas less than 2.5 cm long and strongly urceolate with a ventral, protruding pouch, and with the distance from the base of the corolla to the tip of the pouch considerably longer than the distance from the base to the throat; 2) inflorescences with several flowers; 3) indumentum of plants, if present, usually appressed and not particularly conspicuous; and 4) leaves relatively small (rarely exceeding 15 cm in length and 7 cm in width), 2–3 times as long as wide, and with an acuminate apex, a remotely serrate margin, with most petioles 1–2 cm long, with fewer than 12 secondary veins per side, and with the lower surface bearing more or less conspicuous stomatal clusters.

Most of the stated characters vary beyond the measurements indicated above in some local populations of *G. calcaratus*. Important deviations are the following: 1) corollas may be up to 3.5 cm long in the Colombian department of Valle del Cauca; 2) the ventral pouch is relatively small and hardly protrudes beyond the throat of the corollas both in populations found in eastern (Amazonian) Ecuador and in northwestern Ecuador in the provinces of Carchi and Esmeraldas; 3) a conspicuous pilose to villous indumentum on the corollas of populations found in eastern (Amazonian) Ecuador; 4) obovate leaves with an acute-obtuse apex found in populations in northwestern Ecuador; 5) narrowly falcate leaves approximately 4–5 times longer than wide found in populations in eastern (Amazonian) Ecuador, usually in combination with strongly anisophyllous pairs of leaves; 6) petioles nearly absent due to decurrent leaf bases in lowland populations from the Colombian department of Chocó; and 7) up to 14 secondary veins per side found in populations in the Colombian department of Valle del Cauca (but then the laminas not rugose and pilose between the veins, as in *G. quitensis*).

Gasteranthus calcaratus is one of the most widespread species in the genus, widely distributed in Ecuador and Colombia to Peru. The three subspecies recognized below are geographically disjunct. One subspecies (subsp. *calcaratus*) is found mostly well above 1,000 m elevation in the Central and Western Cordilleras of Colombia, and only found below 1,000 m in the very wet areas of Chocó. The second (subsp. *oncogastrus*) occurs mostly well below 1,000 m elevation in western Ecuador (Fig. 14), and above 1,000 m only in Pichincha province. The third (subsp. *calceolus*) grows mostly below 500 m and rarely above 1,000 m elevation along the eastern foothills of the Andes from southern Colombia to northern Peru.

The three subspecies of *G. calcaratus* are for the most part easily separated. The most distinct is subsp. *calceolus*, which is characterized by strongly anisophyllous leaf pairs with mostly falcate blades, and also in often having corollas with a villous indumentum. *Gasteranthus calcaratus* subsp. *calceolus* is the only taxon in *Gasteranthus* that has strongly anisophyllous leaves, but some specimens from the eastern Andean foothills of Ecuador do not have this character, and can hardly be distinguished vegetatively from the other two subspecies. Subspecies *calcaratus* and *oncogastrus* are primarily set apart by having leaves with 9–14 and 5–8 secondary veins per side, respectively. A few specimens found within the Colombian range of subsp. *calcaratus*, however, may have as few secondary veins per side as the predominantly western Ecuadorian subsp. *oncogastrus*; in these cases corolla length and distribution separate the two subspecies.

The sheet bearing the type specimen of the basionym *Besleria calcarata* Kunth in Paris (*Humboldt & Bonpland 1878*) also includes specimens of *Alloplectus hispidus* (Kunth) Martius. Only the inflorescence and flowers in the center of the sheet are truly *Besleria calcarata*.

6a. Gasteranthus calcaratus subspecies calcaratus.

Besleria tincta C. V. Morton, Contr. U.S. Natl. Herb. 26: 470. 1939. *Gasteranthus tinctus* (C. V. Morton) Wiehler, Selbyana 1: 155. 1975.—TYPE: COLOMBIA. Quindio: Salento, Río Boquía, 1600–1900 m, 27 Jul 1922, *Killip & Hazen 8798* (holotype: PH!; isotypes: GH! NY! US!).

Besleria columbiana var. *arguta* C. V. Morton, Contr. U.S. Natl. Herb. 26: 462. 1939.—TYPE: COLOMBIA. Antioquia: San José, 7000–8000 ft, 28 Mar 1880, *Kalbreyer 1509* (holotype: K!).

Leaves isophyllous; blades maximally 3.5 times as long as wide, usually elliptic and never falcate, with an acuminate apex, margin serrate, rarely subentire, number of secondary veins commonly 9–12, seldom 6–14 per side. Corollas 1.2–3.5 cm long, lacking a conspicuous indumentum. Nectary villous and glandular distally, sparsely glabrescent-glandular basally. Capsule ca. 5 mm long, 7 mm wide; seeds ca. 0.4 mm long, 0.2 mm wide, ellipsoid. Figs. 6A, 12G, H.

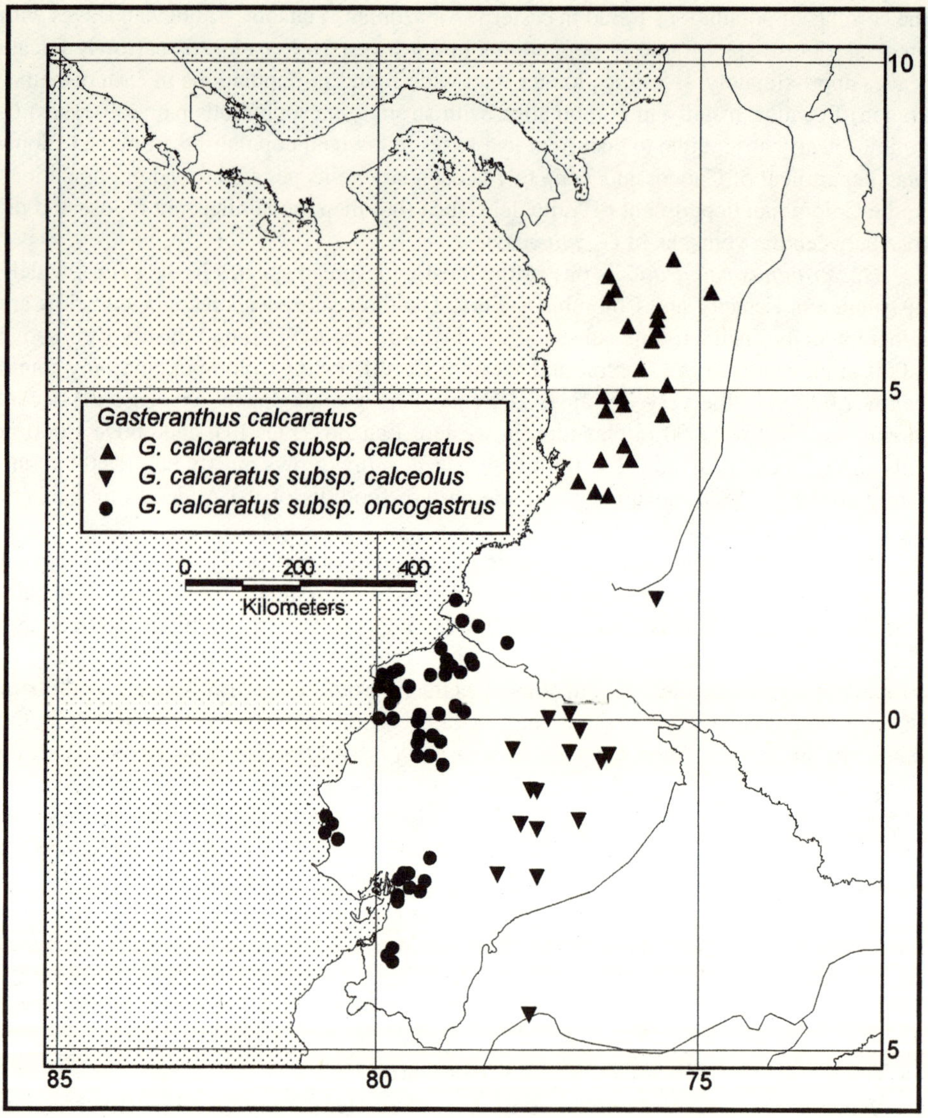

FIG. 18. Distribution of *Gasteranthus calcaratus* subsp. *calcaratus*, subsp. *calceolus*, and subsp. *oncogastrus*.

Phenology. Collected in flower from January to December; a fruiting collection was made in May, but the plants presumably have fruits during much of the year.

Distribution (Fig. 18). Colombia (Antioquia, Caldas, Chocó, Quindio, Risaralda, Valle del Cauca); cloud forest and low montane forest in shade usually near streams; (300–) 1300–2300 (–3100) m.

REPRESENTATIVE SPECIMENS. **Colombia**. ANTIOQUIA: National Park "Las Orquideas," left bank of Río Calles, *Cogollo et al. 2770* (MO), *2275* (MO), *3885* (MO); Mpio. Heliconia, rd Medellín–Heliconia, 16 km NW of San Antonio de Prado, *Callejas et al. 9535* (HUA, US).—CALDAS: Manizales, rd to Neiva, *López s.n.* (US).—CHOCÓ: Ansermanuevo–San José del Palmar rd (border with Valle del Cauca), *Forero et al. 2839* (COL, MO,

US); rd San José de Palmar–Nóvita, close to Curundó, *Forero et al. 3036* (COL, MO, US-2 sheets); trail between Río Negro and Vereda de Playa Rica (at Río Blanco), *Silverstone-Sopkin 1845* (CUVC, MO).—QUINDÍO: Circasia, forest reserve Bremem, *Arbeláez S. et al. 15* (US).—RISARALDA: Mpio. Pereira, old rd to Salento, El Cedral, *Galeano et al. 1987* (COL).—VALLE DEL CAUCA: Río Sanquininí Valley, La Laguna, *Cuatrecasas 15490* (F-2 sheets, US-2 sheets); Mpio. La Elvira, Finca Zingara, *Luteyn et al. 12553* (MO, NY, US); Mpio. El Cairo, Cerro El Inglés, *Silverstone-Sopkin et al. 2980* (F, GB).

Gasteranthus calcaratus subsp. *calcaratus* is endemic to Colombia where it primarily occurs in the Western Cordillera and on the western slopes of the Central Cordillera draining to the Río Cauca (but apparently not on the eastern slopes draining to Río Magdalena). Most collections come from elevations between 1300 and 2300 m, but in the Department of Chocó this subspecies is found as low as 300 m.

6b. Gasteranthus calcaratus subspecies **calceolus** (Fritsch) L. E. Skog & L. P. Kvist, stat. nov. *Besleria calceolus* Fritsch, Repert. Spec. Nov. Regni Veg. 18: 12. 1922. *Gasteranthus calceolus* (Fritsch) Wiehler, Selbyana 1: 154. 1975.—TYPE: ECUADOR. Pastaza?: Montañas de Canelos, 1857, *Spruce 5069* (holotype: W!; isotypes: BM! C! CGE! E-2 sheets! G-3 sheets! GH! GOET! K-3 sheets! MPU! P! US! W-2 sheets!).

Besleria caligula C. V. Morton, Contr. U.S. Natl. Herb. 38: 148. 1968. *Gasteranthus caligula* (C. V. Morton) Wiehler, Selbyana 1: 154. 1975.—TYPE: PERU. Loreto: Alto Amazonas, near Río Marañón just above Pongo de Manseriche, 250–350 m, 17 Oct 1962, *Wurdack 2277* (holotype: US!; isotypes: K! NY! USM!).

Leaves mostly strongly anisophyllous but sometimes subequal; blades up to 6 times as long as wide, mostly narrowly falcate to narrowly oblanceolate, occasionally ovate-lanceolate with an acuminate apex, margin subentire to remotely serrate, number of secondary veins per side 8–12. Corollas 1.2–2.5 cm long, often with a conspicuous villous indumentum on calyx and corolla. Nectary glabrous. Capsule ca. 4 mm long, 6 mm wide; seeds ca. 0.3 mm long, 0.2 mm wide, ovoid. Figs. 6B, 12B, C, D.

Phenology. Collected in flower from January to July, in September and October; in fruit during July.

Distribution (Fig. 18). Colombia (Caquetá), Ecuador (Morona-Santiago, Napo, Pastaza, Sucumbíos), and Peru (Loreto); undisturbed dense rainforest, especially premontane and montane wet forest in shady wet places usually along streams; 200–950 (–1700) m.

REPRESENTATIVE SPECIMENS. **Colombia**. CAQUETÁ: Florencia–Guadalupe rd, Km 21, *Londoño & Kvist 128* (AAU, TULV, US). **Ecuador**. MORONA-SANTIAGO: rd Limón–Macas, 95 km NE of Limón, *Bohlin et al. 1478* (GB).—NAPO: Yuca rd, 3 km from Auca oil-field rd, *Brandbyge et al. 30292* (QCA, US-2 sheets); Archidona Cantón, Reserva Ecológica Antisana, comunidad Shamato, entrada por Km 21–Shamato, Camino Sardinas–Shamato, *Clark et al. 5325* (AAU, COL, MO, QCNE, SRP, US); left bank of Río Napo, Campanacocha–Dayona rd, *Jaramillo & Coello 3741* (MO, QCA-2 sheets); Añangu, NW corner of Yasuní National Park, *Korning & Thomsen 47002* (AAU, QCA), *47108* (AAU, QCA).—PASTAZA: 15 km N of Puerto Sarayacu, Río Chullana, *Lugo S. 4195* (AAU, F, GB, NY, SEL); 16 km NW of Sarayacu, Pacayacu at Río Bobonaza, *Lugo S. 5254* (GB, US).—SUCUMBÍOS: rd Lago Agrio–El Chaco, *Lugo S. 3500* (AAU, F, GB, QCA, SEL, US).

Gasteranthus calcaratus subsp. *calceolus* is known from the Colombian department of Caquetá in the north to the Peruvian department of Loreto in the south; however, it has only been collected once in each of these two countries. All the remaining collections come from the intervening Ecuadorian Amazon region, where it is locally common in the province of Napo. Nearly all collections come from below 500 m elevation, and the subspecies has rarely been collected above 1,000 m elevation.

6c. Gasteranthus calcaratus subspecies **oncogastrus** (Hanstein) L. E. Skog & L. P. Kvist, stat. nov. *Besleria oncogastra* Hanstein, Linnaea 34: 335. 1865. *Gasteranthus oncogastrus* (Hanstein) Wiehler, Selbyana 1: 155. 1975.—TYPE: ECUADOR. Guayas: Guayaquil, *Ruiz s.n.* (lectotype, designated Morton, 1939: B, destroyed, fragment at F!; the isolectotype at US is here designated as a new lectotype).

Leaves isophyllous; blades maximally 3.5 times as long as wide, mostly elliptic-oblanceolate with an acuminate apex, occasionally ovate to obovate with an acute-obtuse apex, margin usually remotely serrate, number of secondary veins 5–8 per side. Corollas 1.2–2.5 cm long, lacking a conspicuous indumentum. Nectary densely pubescent. Capsule ca. 4 mm long, 9 mm wide; seeds ca 0.4 mm long, 0.2 mm wide, ellipsoid. Figs. 6C, 12I, J.

Phenology. Collected in flower from January to December; in fruit in March, April, September, and November.

Distribution (Fig. 18). Colombia (Nariño) and Ecuador (Azuay, Bolívar, Cañar, Carchi, El Oro, Esmeraldas, Guayas, Los Ríos, Manabí, Pichincha); wet mature forest, lower montane forest, and coastal lowland seasonal forest, in shade often near streams; 0–900 (–2050) m.

REPRESENTATIVE SPECIMENS. **Colombia**. NARIÑO: Mpio. Tumaco, Llorente, resguardo Alto Albí, El Trampal, *González 271* (US-2 sheets). **Ecuador**. AZUAY: Cuenca, Molleturo Mullopungo protected forest, near Manta Real, *Clark et al. 2484* (MO, QCNE, SRP, US).—BOLÍVAR: Hda. Changuil, Nuevo Mundo, *Cornejo & Bonifaz 4322* (US).—CAÑAR: Azogues–El Triumfo rd, 5 km W of La Delicia, *Croat 50890* (MO, US).—CARCHI: Maldonado, Awa ethnic reserve, Sabalera, *Aulestia et al. 691* (MO, QCNE).—EL ORO: rd from Piñas to Santa Rosa, *Dodson et al. 8908* (MO, QCNE, SEL).—ESMERALDAS: Río Zapallo Grande (tributary of Río Cayapa), *Barfod 41000* (AAU, MO, QCA, US); Reserva Ecológica Mache–Chindul, Comunidad Caña Braval, Río Viche, *Clark 4712* (AAU, COL, MO, QCA, QCNE, US); Esmeraldas–Quinindé rd, 40–50 km SE of Esmeraldas, *Harling & Andersson 16728* (GB, QCA, US); Río Cayapa, Zapallo Grande, *Kvist & Asanza 40327* (AAU, NY, QCA, QCNE), *40734* (AAU, QCA, QCNE); Bilsa Biological Station, *Mendoza-T. et al. 577* (AAU, QCA, QCNE, US).—GUAYAS: hills E of Naranjal–Machala rd, 13 km S of Naranjal, *Harling & Andersson 19332* (AAU, GB, QCA, US).—LOS RÍOS: Quevedo–Santo Domingo, Km 56, Río Palenque Biological Station, *McMahon & Dodson 4281* (F, QCA, SEL-4 sheets).—MANABÍ: Machalilla National Park, S of San Sebastián, *Øllgaard et al. 100815* (AAU).—PICHINCHA: Santo Domingo–Puerto Limón rd, Km 23, Colorado community "Congóma Grande," *Kvist & Holm-Nielsen 40233* (AAU, QCA); Santo Domingo–Quito rd, Tinalandia, *Mendoza-T. et al. 600* (AAU, B, BM, COL, GB, MO, NY, QCA, QCNE, SEL, U, US); Santo Domingo–Quinindé rd, Km 41, protected forest "La Perla," *Zak et al. 5369* (QCA, QCNE), *5538* (QCA, QCNE), *5551* (QCA, QCNE).

See discussion under *Gasteranthus lateralis* (no. 19) for a comparison of that species with *G. calcaratus* subsp. *oncogastrus*.

7. Gasteranthus carinatus Wiehler, Selbyana 2: 77, pl. 23D. 1977.—TYPE: ECUADOR. Los Ríos or Pichincha: Montañas de Ila, 12 km E of Río Palenque Science Center (located at Km 56 on road between Quevedo and Santo Domingo), 650 m, 9 Apr 1977, *Madison 3802* (holotype: SEL!).

Subshrubs to shrubs; stems erect, 1–2 (–5) m tall, to 8 mm in diameter (so noted on herbarium labels, but the taller plants probably larger at base), glabrous, internodes

1–4 cm long, terete to quadrangular, with wrinkled cortex. Leaves opposite, isophyllous; blades 12–32 cm long, 7–14 cm wide, ovate to elliptic, base cuneate to obtuse, apex acuminate, margin serrate to subentire, usually somewhat rugose (but appearing flat in herbarium specimens), adaxially dark green, glabrous, abaxially lighter green, glabrous, lateral veins 10–17 per side, stomatal clusters not conspicuous; petioles 1–6 cm long, glabrous. Inflorescences usually congested cymes of 10–30 (or more) flowers, glabrous; peduncles 4–14 cm long; pedicels 4–8 mm long. Calyx apparently green to yellow, glabrous, lobes nearly free to base, subequal, each of the lobes 5–10 mm long, 4–8 mm wide, ovate with acute to obtuse apex, entire, dorsal lobe directed backwards surrounding spur, ventral lobes directed forward, lateral lobes directed upward. Corolla 1.7–2.5 cm long, subventricose to funnelform with small limb, often with dorsal longitudinal keel, tube 1.2–1.7 cm long, widened from 2–4 mm at base to 3–6 mm in diameter at widest point, orange, outside glabrous, inside glabrous, markings not seen; spur 3–6 mm long, 2–3 mm wide, apex blunt, slightly bent downward; throat 3–6 mm in diameter; limb orange, lobes 2–3 mm long, 2–3 mm wide, subequal, rotund, glabrous. Filaments 7–9 mm long, adnate to base of corolla tube for ca. 2 mm, glabrous; anthers ca. 1.5 mm long, 1 mm wide. Nectary dorsal, bilobed, ca. 1 mm high, glabrous. Ovary ca. 3 mm long, 3 mm in diameter, glabrous; style ca. 8 mm long, glabrous; stigma stomatomorphic. Capsule ca. 3 mm long, ca. 7 mm wide, shape unknown, bivalved; seeds ca. 0.3 mm long, 0.1 mm wide, elliptic, puberulent. Fig. 8A.

Phenology. Collected in flower in February, April, July, August, and October to December; in fruit in August.

Distribution (Fig. 19). Ecuador (Azuay, El Oro, Los Ríos, and Pichincha); wet forest, in shady wet areas; 400–900 (–2000) m.

ADDITIONAL SPECIMENS EXAMINED. **Ecuador**. AZUAY: Km 8 on rd from C.P. Enrique to Nuevo Pueblo, *Klitgård 67038* (AAU).—EL ORO: New Piñas–Machala rd, 10 km W of Piñas, *Dodson et al. 8446* (SEL); Km 19, rd from Piñas to Santa Rosa, *Dodson et al. 8909 p.p.* (QCNE, SEL [the MO-duplicate of the same number is *G. calcaratus* subsp. *oncogastrus*]); between Huertas and Malvas, *Escobar 769* (HUA, QCA, SEL); trail from Sambotambo, following headwaters of Río Moro Moro, S to Buenaventura at and along hwy to Portovelo, *Steyermark 54215* (F, US).—LOS RÍOS OR PICHINCHA: Montañas de Ila, El Centinela, *Dodson 7297* (AAU, MO, SEL); Centinela, *Dodson et al. 7515* (F, MO, SEL); Centinela, *van der Werff et al. 12387* (QCNE).

Gasteranthus carinatus is a somewhat shrubby species, characterized by the following combination of characters: 1) small funnelform to subventricose orange corollas; 2) all parts of plants glabrous; 3) epidermis of stems wrinkled, marked by more or less irregular longitudinal ridges; and 4) large leaves with many secondary veins (usually more than 12 per side).

The most similar species may be the also shrubby, wrinkle-barked, nearly glabrous, and large-leaved *G. imbricans* from Panama and Costa Rica. The main differences between the two taxa are in the flowers (Fig. 8). *Gasteranthus imbricans* (Fig. 8C) differs from *G. carinatus* (Fig. 8A) by having the spur bent upward rather than slightly downward, by having wider and more imbricate calyx lobes, and by having corollas with a wider throat that bear glandular hairs. Another species with similar flowers is *G. columbianus* from the western montane forests of northern Ecuador and southern Colombia. This species differs from *G. carinatus* by having strigillose lower leaf surfaces, inflorescences, and calyces, and smaller leaves with fewer veins and usually conspicuous stomatal clusters; it also has corollas with a glandular-hairy throat (Fig. 8B). In addition

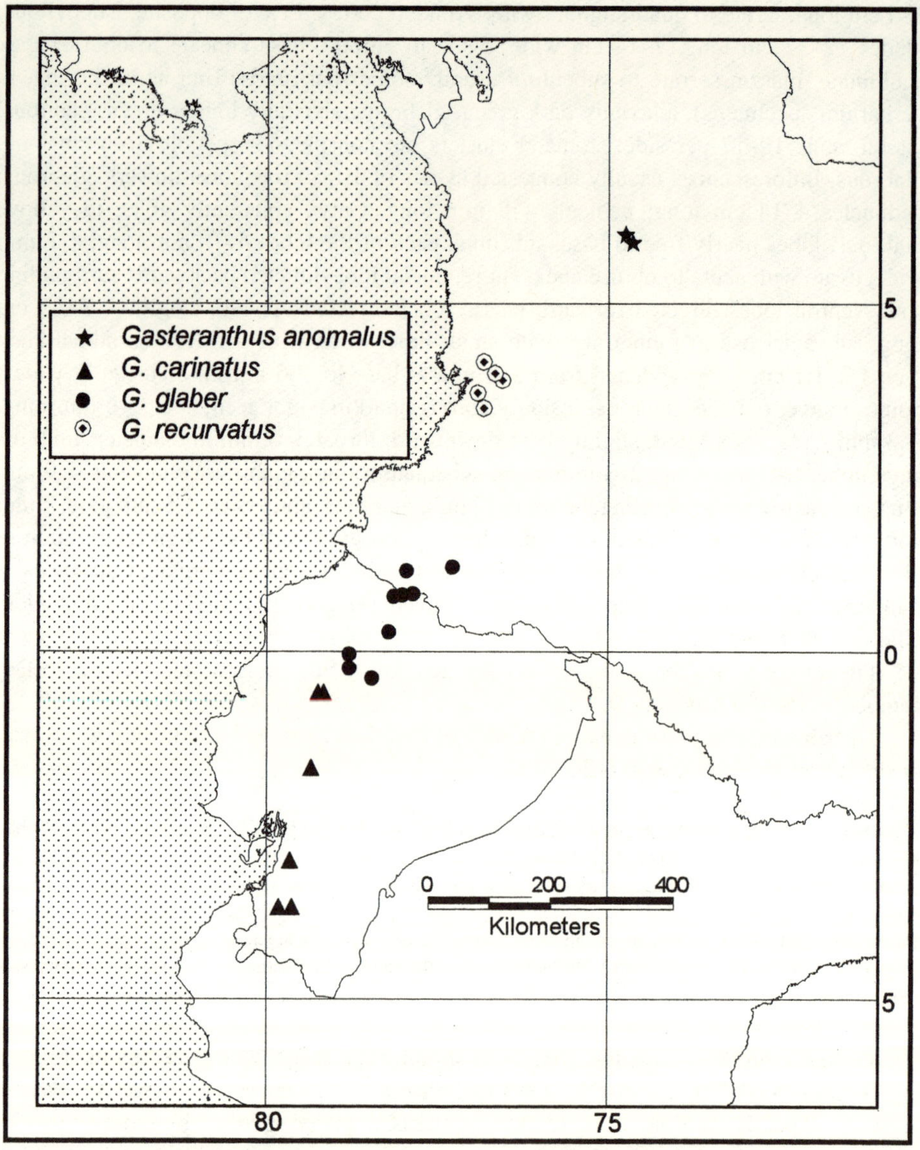

FIG. 19. Distribution of *Gasteranthus anomalus*, *G. carinatus*, *G. glaber*, and *G. recurvatus*.

to *G. carinatus* only *G. glaber* is nearly glabrous, but this species has urceolate corollas with a blackish limb (see *G. glaber*, no. 15).

Wiehler (1977) named *G. carinatus* for its corollas with a distinctive dorsal keel of unknown function (Fig. 8A), but it is reduced or absent particularly in the collections from the provinces of Azuay and El Oro. These specimens also have somewhat smaller flowers, but are otherwise similar to the collections from Los Ríos (and adjacent Pichincha). No other species of *Gasteranthus* apparently has this keel.

8. Gasteranthus columbianus (C. V. Morton) Wiehler, Selbyana 1: 155. 1975. *Besleria columbiana* C. V. Morton, Contr. U.S. Natl. Herb. 26: 462. 1939.—TYPE: COLOMBIA. Nariño: West Andes of Tuquerres, above Pipulquer, 1500–1700 m, *Lehmann 5157* (holotype: K!; isotype: K!).

Herbs or subshrubs; stems erect, to 1 m tall, to 6 mm in diameter, distally strigillose, internodes 1–4 cm long, terete to quadrangular. Leaves opposite, isophyllous; blades 6–12 cm long, 2–5 cm wide, ovate to elliptic or oblanceolate, base cuneate, occasionally acute, apex acute, margin serrate, flat, adaxially dull green, mostly nearly glabrous, occasionally appressed pilose-strigillose, abaxially lighter green, strigillose, particularly along veins, lateral veins 8–11 per side, stomata in conspicuous yellow-green clusters; petioles 1–5 cm long, strigillose. Inflorescences open to relatively congested cymes of 2–10 flowers, strigillose; peduncles nearly vertical, 4–10 cm long; pedicels 3–8 mm long. Calyx green, outside strigillose, lobes nearly free to base, subequal, each of the lobes 5–7 mm long, 3–6 mm wide, ovate with acute apex, subentire, dorsal lobe directed backward, surrounding spur, ventral lobes directed forward, lateral lobes directed upward. Corolla 1.3–1.9 cm long, tube 0.8–1.2 cm long, funnelform with a relatively small limb, widened from 2–4 mm at base to 6–11 mm in diameter at widest point, yellow-orange-red, outside puberulent, inside glandular-hairy, particularly near throat, markings not seen; spur 3–5 mm long, 3–6 mm wide, blunt, straight; throat 5–9 mm in diameter; limb orange-red, lobes 1.5–3 mm long, 2–4 mm wide, subequal, rotund, glabrous. Filaments 6–11 mm long, adnate to base of corolla tube for 3–5 mm, glabrous; anthers ca. 2 mm long, 2 mm wide. Nectary thickened dorsally, annular, to 0.8 mm high, sparsely pilose distally, glabrescent basally. Ovary ca. 3 mm long, 2 mm in diameter, glabrous; style ca. 4 mm long, glabrous; stigma stomatomorphic. Capsule ca. 4 mm long, 7 mm wide, flattened, bivalved, secondarily splitting into 4 valves; seeds ca. 0.3 mm long, 0.2 mm wide, ovoid, papillate. Figs. 8B, 13C, D.

Phenology. Collected in flower from November to May and in July; in fruit during March, April, November, and December.

Distribution (Fig. 20). Colombia (Nariño) and Ecuador (Carchi, Imbabura); cloud forests or wet montane forests in shady areas, especially along streams; (780–) 1300–2200 (–2900) m.

ADDITIONAL SPECIMENS EXAMINED. **Colombia**. NARIÑO: La Planada reserve, 7 km from Chucunés, *Benavides 8840* (US); La Planada reserve, Quebradas El Mar–La Calladita, *Benavides 9621* (US); La Planada, Isla de Los Osos, *Benavides 9775* (MO, US); trail Pialapi–La Planada, *Benavides 10162* (US); Corregimiento Altaquer, Vereda El Barro, natural reserve "Río Ñambi," *Betancur et al. 4748* (COL), *4885* (COL, US); La Planada, Tuquerres–Ricaurte rd, 7 km above Chucunés, *Croat 69606A* (US), *Croat 71478* (MO); trail from La Planada to Pielapi, *Gentry et al. 63588* (US); La Planada, 7 km from Chucunés, *Giraldo 151* (HUA); Mpio. Ricuarte, Finca El Braque, secc. Acantayoc, vereda San Isidro, *Restrepo 660* (US); Mpio. Ricuarte, border Marcos, *Restrepo 746* (US); La Planada, *Snow 22* (K); La Planada, *Stiles 570* (COL-2 sheets). **Ecuador**. CARCHI: 45 km below Maldonado, *Madison & Besse 7059* (SEL); above Río Verde, mountain ENE of Quindi's finca, *Hoover 2283* (QCA); Gualpi Chico, *Hoover et al. 2468* (QCA, US).—IMBABURA: Cotocachi, Tablachupa, NW of the Laguna de Cuicocha, *Clark 1746* (COL, MO, QCNE, US); Cotocachi Cantón, Tablachupa, Apuela, *Gudiño & Cuamacás 1954* (QCNE, US); SW slopes of Cotocachi volcán, upper Intag valley, *Molau et al. 2650* (GB, QCA); Cotocachi cantón, rd to Apuela, Tabla Chupa, *Palacios & Gudiño 9863* (QCNE).

Gasteranthus columbianus has small (1.3–1.9 cm long), yellow-orange or red, funnelform corollas (Fig. 8B), mostly in relatively few-flowered inflorescences, which appear terminal owing to the often nearly vertical orientation of the peduncles. The most similar

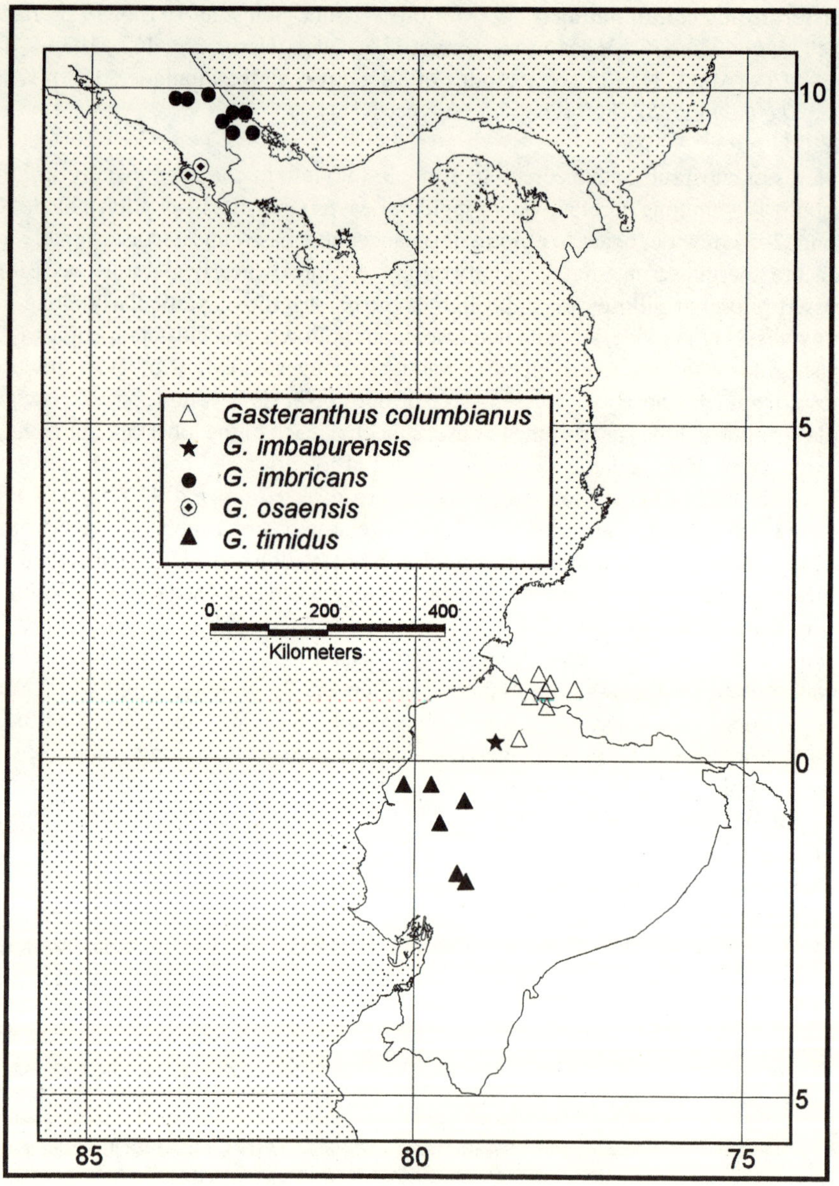

FIG. 20. Distribution of *Gasteranthus columbianus*, *G. imbaburensis*, *G. imbricans*, *G. osaensis*, and *G. timidus*.

corollas may be found in *G. carinatus* also of western Ecuador but more to the south, but this species is totally glabrous, whereas *G. columbianus* mostly has a strigillose indumentum. Also, *G. carinatus* usually has inflorescences with more rather than fewer than six flowers; it is a shrubbier species with larger leaves and a conspicuously wrinkled bark. Specimens of *G. columbianus* have mostly been confused with the two sympatric species

G. corallinus (Fig. 10D) and *G. calcaratus* (Fig. 6), but both of these species differ from *G. columbianus* by having urceolate rather than funnelform corollas. The former is easily distinguished even without corollas by its stems with a conspicuous and often villous indumentum, and by its calyx lobes with an incised-serrate rather than an entire margin. In contrast, it may be difficult to distinguish *G. columbianus* from *G. calcaratus* without fully developed corollas. Neither of them has a particularly conspicuous indumentum, but a magnifying glass will disclose that the former has a mostly strigillose indumentum, and the latter a more or less arachnoid indumentum. Some populations of *G. calcaratus* subsp. *oncogastrus* from the Ecuadorian provinces of Esmeraldas and Carchi have urceolate corollas (Fig. 6C) with unusually poorly developed pouches (see *G. calcaratus* subsp. *oncogastrus*, no. 6c), and may thus resemble *G. columbianus*.

The few collections of *G. columbianus* from Imbabura in Ecuador are those that come from the highest elevations (ca. 2900 m); these sites are among the highest localities of any species of *Gasteranthus*.

9. Gasteranthus corallinus (Fritsch) Wiehler, Selbyana 1: 154. 1975. *Besleria corallina* Fritsch, Repert. Spec. Nov. Regni Veg. 18: 13. 1922.—TYPE: PERU. San Martín: Cerro de Ponasa, 1200 m, Feb 1903, *Ule 6671* (holotype: B, destroyed, photo: US!; lectotype, here designated: HBG!).

Besleria corallinoides Fritsch, Repert. Spec. Nov. Regni Veg. 18: 11. 1922. *Gasteranthus corallinoides* (Fritsch) Wiehler, Selbyana 1: 154. 1975.—TYPE: ECUADOR. Pichincha: Río Pilatón, 800 m, Oct 1882, *Sodiro 119/55* (holotype: B, destroyed, photo: US!; lectotype, here designated: QPLS, photo: US!).

Besleria sylvarum C. V. Morton, Contr. U.S. Natl. Herb. 26: 459. 1939. *Gasteranthus sylvarum* (C. V. Morton) Wiehler, Selbyana 1: 155. 1975.—TYPE: COLOMBIA. Cauca: District of El Tambo, La Costa, 1000 m, 2 Aug 1936, *Sneidern 957* (holotype: S!).

Besleria crenata C. V. Morton, Contr. U.S. Natl. Herb. 26: 460. 1939. *Gasteranthus crenatus* (C. V. Morton) Wiehler, Selbyana 1: 154. 1975.—TYPE: COLOMBIA. Nariño: San Pablo, May 1876, *André 3329* (holotype: K!).

Herbs to subshrubs or rarely shrubs; stems erect, to 1.5 m tall (but one collection noted to be 4 m tall), to 1 cm in diameter, glabrous to densely arachnoid-tomentose or villous toward the apex, trichomes less than 2 mm long, internodes 1–5 cm long, mostly quadrangular. Leaves opposite, subisophyllous; blades 7–30 cm long, 2–16 cm wide, oblong or obovate, base cuneate to acute, often oblique, apex acute, margin serrate, dentate to nearly entire, flat, adaxially dark green, glabrous or very sparsely pilose, abaxially lighter green, appressed pilose to tomentose along the prominent veins, lateral veins 8–12 (–16) per side, stomatal clusters not conspicuous; petioles mostly less than 1 cm but occasionally up to 4 cm long, arachnoid to tomentose or glabrous. Inflorescences congested cymes of 5–20 flowers, more or less tomentose or arachnoid, but pedicels often glabrous; peduncles 4–13 cm long; pedicels short, rarely exceeding 1 cm. Calyx green, outside mostly nearly glabrous or pilose, lobes basally connate 1–2 mm, subequal, each of the lobes 5–8 mm long, 4–7 mm wide, ovate or deltate with acute apex, incised to serrate, dorsal lobe usually reflexed, dorsal lobe also directed backward below spur, ventral lobes directed forward but usually reflexed, lateral lobes directed forward but usually reflexed. Corolla 0.9–2.2 cm long, urceolate with a small limb and a ventral pouch, sometimes approaching ventricose, tube 0.7–1.5 cm long, ca. 1 mm in diameter, widened from 2–4 mm at base to maximally 5–9 mm in diameter at widest point, red, scarlet, or orange, outside appressed-puberulent, inside glabrous to puberulent, markings none; spur 1–3 mm long, to 3 mm wide, blunt, straight; throat 2–4 mm in diameter; limb red, lobes 1–2 mm long, 1–3 mm wide, equal, rotund, glabrous. Filaments 6–10 mm long, adnate to base of corolla tube for 2–3 mm, glabrous; anthers ca. 1.2 mm long, 0.8 mm wide. Nectary dorsal or thickened dorsally, bilobed or semiannular, to 1 mm high, pilose-pubescent distally, glabrescent toward the base. Ovary 3–4 mm long, 2–4 mm in diameter, minutely puberulent or glabrous; style ca. 6 mm long, glabrous; stigma stomatomorphic to weakly bilobed. Capsule ca. 3 mm long, ca. 7 mm wide, subglobose, bivalved, partly splitting into 4 valves, glabrous, white at maturity; seeds ca. 0.4 mm long, 0.3 mm wide, broadly ellipsoid, papillate. Figs. 10D, 21.

Phenology. Collected in flower from January to December; in fruit from January to October.

Distribution (Fig. 22). Colombia (Amazonas, Cauca, Nariño, Putumayo), Ecuador (Bolívar, Carchi, Esmeraldas, Imbabura, Morona-Santiago, Napo, Pastaza, Pichincha, Sucumbíos, Zamora-Chinchipe), and Peru (Amazonas, Loreto, San Martín); dense wet lowland to montane forests, and common along shaded streams; 70–2000 m.

REPRESENTATIVE SPECIMENS. **Colombia**. AMAZONAS: Mpio. Leticia, Parque Nacional Natural Amacayacu, Quebrada de Agua Pudre, *Pipoly et al. 16445* (MO); Loretoyacu River, *Schultes & Black 8329* (GH), *8329A* (US).—CAUCA: W of Tambo, *Haught 5214* (COL, US).—NARIÑO: along trail above San Juan River, *Hoover 1034* (QCA, US); El Palmar, *Sneidern 4521* (A, F, NY, US); Mpio. Ricaurte, Reserva Natural La Planada, *Stiles 574* (COL).—PUTUMAYO: Quebrada de la Hormiga, Río San Miguel, *Cuatrecasas 11088* (F, US); between Mocoa and Sachamates, *Cuatrecasas 11409* (F, US). **Ecuador**. BOLÍVAR: Tablas de Telimbela, *Acosta Solís 6949* (US).—CARCHI: Maldonado, Parroquia Tobar Donoso, Awá Ethnic Reserve, Sabalera, *Aulestia et al. 829* (US); Espejo, Bosque Protector Mirador de Golondrinas, between Las Juntas and La Cabaña de Corazón, *Clark et al. 2414* (MO, QCNE, SRP, US); Gualpi Chico, Awá Encampment, *Hoover et al. 2470* (MO, QCA).—ESMERALDAS: environs of Lita, on the Ibarra–San Lorenzo R.R., *Madison et al. 5133* (QCA, SEL); Parroquia Concepción, Playa Rica, *Mexia 8438* (F, GB, GH, K, MO, NY, US).—IMBABURA: Lita, *Acosta Solís 12220* (F).—MORONA-SANTIAGO: 10 km NW of General Proaño on rd to Alshi (=9 de Octubre), *Dorr & Valdespino 6329* (QCA); Macas–Sucua rd, Km 10, *Kvist 60454* (US).—NAPO: Cantón Archidona, slopes of Volcán Sumaco, Hollín–Loreto rd, Km 31, *Alvarado 250* (QCNE, US-2 sheets); Jatún Sacha Biological Reserve, Río Napo, 8 km below Misahuallí, *Cerón 805* (MO, NY, QCA, QCNE); Nuevo Rocafuerte, Río Braga, *Jaramillo & Coello 4643* (NY, QCA); rd from Tena to Puyo, 4.7 km S of Puerto Napo, *Thomas & Rios 6665* (QCA).—PASTAZA: Curaray (Jesús Pitishka), *Harling & Andersson 17406* (F, GB, US); trail towards Colonia 24 de Mayo, 2.5 km W on rd departing main Puyo–Tena rd at Km 9, *Stein & Tucker 3110* (US).—PICHINCHA: ENDESA Reserve, ca. 6 km WNW of P. Vicente Maldonado, *Harling & Andersson 23355* (GB, QCA); ENDESA Reserve, Río Silanche, Km 113 on Quito–Pto. Quito hwy, *Jaramillo 6419* (AAU, GB-2 sheets, NY, QCA).—SUCUMBÍOS: Comunidad Sanisla, N side of Río Napo, 15 km E of Pompeya, *Clark 4474* (QCNE, US); rd Lago Agrio–El Conejo, between Lago Agrio and Proyecto San Miguel, *Harling & Andersson 16628* (GB, US); environs of Limoncocha, *Madison et al. 5317* (F, QCA, SEL, US). **Peru**. AMAZONAS: Prov. Bagua, Distr. Imaza, Kampaensa, NW of Marañón RENOM, *Quipiscoa S. 338* (MO); valley of Río Santiago, Quebrada Caterpiza, 2–3 km above the Caterpiza community, *Tunqui 715* (US).—LORETO: Prov. Maynas, trail between Indiana (on Río Amazonas) and Mazán (on Río Napo), *Gentry et al. 25445* (F, IBE, NY, US); above Pongo de Manseriche, right side of mouth of Río Santiago, *Mexia 6226* (F, GB-2 sheets, NY, US).—SAN MARTÍN: Prov. Mariscal Cáceres, Dtto. Tocache Nuevo, Santa Rosa de Mishollo, *Schunke V. 6822* (MO).

Gasteranthus corallinus is recognized by its densely congested inflorescences of small (0.9–2.2 cm long), urceolate, and very firm red or orange corollas, each with a ventral pouch. The corolla pouch, however, does not extend beyond the narrow (2–4 mm wide) corolla throat (Figs. 10D, 21C). In addition, the inflorescences appear terminal owing to a mostly nearly vertical orientation of the peduncles. The throat of the corollas

FIG. 21. *Gasteranthus corallinus*. A. Habit. B. Inset: section of abaxial leaf surface. C. Corolla. D. Corolla opened to show stamens. E. Calyx. F. Calyx, pistil, and disk. G. Fruit. H. Seeds. (Based on: A, B, *Croat 73116* C, D, *Cerón et al. 4592*; E–H, *Brandbyge et al. 30019*.)

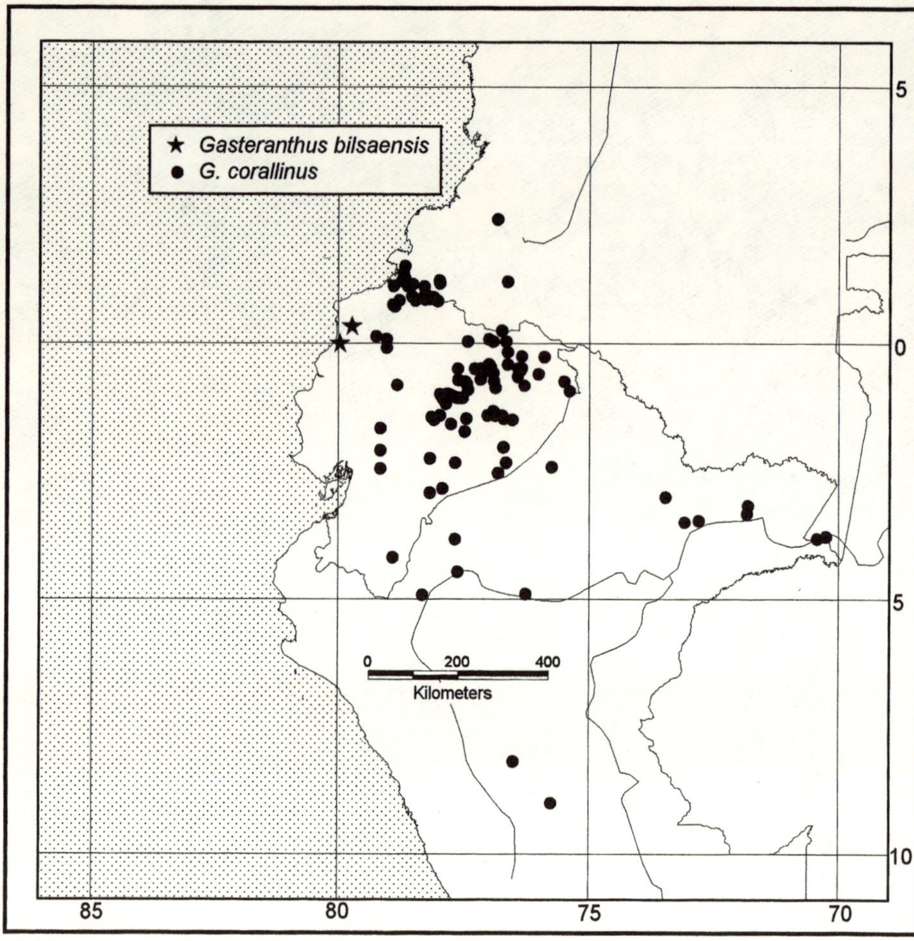

FIG. 22. Distribution of *Gasteranthus bilsaensis* and *G. corallinus*.

also is directed more vertically here than in other species of *Gasteranthus*. Other characteristics of the species include the calyx lobes, which are mostly reflexed and have an incised-serrate margin, the stems and inflorescences usually with a conspicuous indumentum, and the absence of conspicuous stomatal clusters on the abaxial surface of the blades.

The species most similar to *G. corallinus* is *G. bilsaensis*, also from the Ecuadorian province of Esmeraldas. This latter species has the same type of flowers and inflorescences, although usually with much shorter peduncles, but *G. bilsaensis* is easily set apart by its strongly rugose leaves (Fig. 17). The only other species that may have similarly shaped small ventricose-urceolate corollas is *G. glaber* (Fig. 27), found on the western Andean slopes in Ecuador and Colombia, but this species differs among other characters from *G. corallinus* by being nearly glabrous and by having corollas with a blackish limb.

Gasteranthus corallinus is among the most common and widely distributed species in the genus. It occurs both east and west of the Andes. To the east, it ranges along the eastern slopes of the Andes from the department of San Martín in Peru northward through

Ecuador to Putumayo in Colombia. West of the Andes *G. corallinus* ranges from the Ecuadorian province of Bolívar northward to the Colombian department of Cauca. On the eastern Andean slopes *G. corallinus* has been collected at sites up to 1600 m elevation, but most collections come from elevations below 500 m, particularly in the Ecuadorian Amazon region where *G. corallinus* is common. The species also occurs as far away from the Andes as near Leticia in the Colombian department of Amazonas at only 80 m. Most of the western collections come from elevations between 500 and 1000 m, but the relatively few Colombian collections come from elevations between 1,000 and up to nearly 2000 m. In the Ecuadorian province of Esmeraldas *G. corallinus* occurs as low as near sea level (Fig. 14).

Although *G. corallinus* is a widely distributed and common species (Fig. 22), the various populations throughout the range are relatively similar. Plants in western Ecuador and Colombia generally are taller (often to 1–2 m) than those from the Amazon lowland (mostly up to 50 cm), but there are apparently no consistent differences among the populations.

10. Gasteranthus crispus (Mansfeld) Wiehler, Selbyana 1: 154. 1975. *Halphophyllum crispum* Mansfeld, Repert. Spec. Nov. Regni Veg. 41: 145. 1936.—TYPE: ECUADOR. Pichincha: San Carlos de los Colorados, 150 m, 5 Feb 1936, *Schultze-Rhonhof 2031* (holotype: B, destroyed).—ECUADOR. Esmeraldas: Mache mountains, 35 km W of Quinindé, forest at Bilsa Biological Station, 550–650 m, 25 Oct 1995, *Mendoza-T. et al. 569* (neotype, here designated: QCNE!; isoneotypes: AAU! COL! K! MO! QCA! US!).

Herbs; stems decumbent to erect, often appressed to ground, to 30 cm tall, to 5 mm in diameter, glabrous to more or less tomentose distally, internodes 1–3 cm long, usually terete. Leaves opposite, mostly anisophyllous; blades 4–14 cm long, 3–10 cm wide, obovate, nearly orbicular, base obtuse, occasionally acute, oblique, apex obtuse, margin crenate, serrulate, or subentire, often undulate, flat to rugose along secondary veins, adaxially shining dark green, glabrous, abaxially lighter green with conspicuous darker veins, villous except at stomatal clusters, lateral veins 5–7 per side, stomatal clusters very conspicuous; petioles 0.5–3.5 cm long, glabrous to pilose or villous. Inflorescences open cymes of 1–3 flowers, pilose; peduncles 4–7 cm long; pedicels 0.7–2 cm long. Calyx light green, outside arachnoid to tomentose, inside velutinous, lobes basally connate 1–3 mm, unequal, each of the lobes broadly ovate with obtuse apex, dorsal lobe keeled, 6–11 mm long, 5–13 mm wide, lobes approximately equally long but dorsal lobe much wider than particularly the ventral lobes, crenulate to erose or subentire, dorsal lobe directed backward below spur, ventral and lateral lobes directed forward and appressed to corolla. Corolla 3–3.8 cm long, urceolate with a small limb and a ventral pouch, tube (defined as distance from base to throat) 2.0–2.5 cm long but distance from base to tip of pouch 2.5–3.2 cm, widened from 4–6 mm at base to maximally 1.3–2.0 cm in diameter at widest point, bright yellow-orange to red, outside puberulent to tomentose, inside puberulent, markings none; spur 4–6 mm long, to 4 mm wide, blunt, straight; throat 2–4 mm in diameter; limb red, lobes 1–2 mm long, 2–3 mm wide, subequal, rotund to emarginate, puberulent. Filaments 13–18 mm long, adnate to base of corolla tube for 3–4 mm, puberulent; anthers ca. 2 mm long, 2 mm wide. Nectary dorsal, bilobed, ca. 1.5 mm high, directed backward into spur, puberulent to pilose. Ovary ca. 5 mm long, 5 mm in diameter,

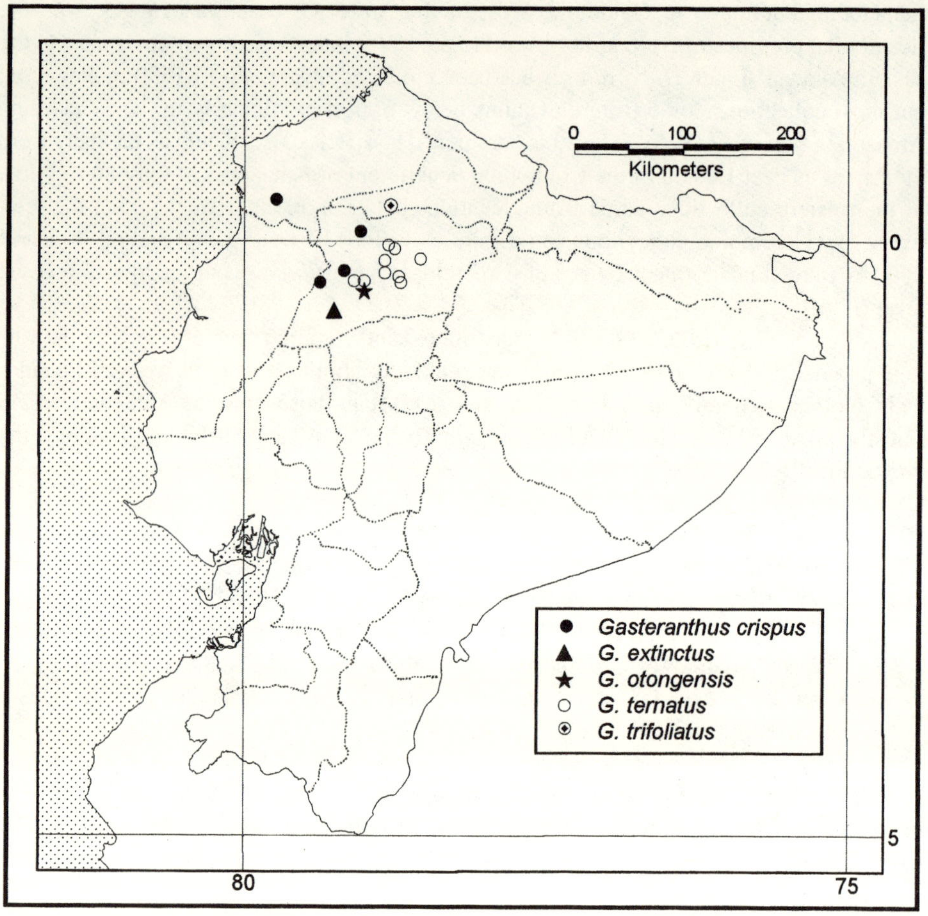

FIG. 23. Distribution of *Gasteranthus crispus*, *G. extinctus*, *G. otongensis*, *G. ternatus*, and *G. trifoliatus*.

puberulent; style 9–11 mm long, puberulent; stigma bilobed. Capsule ca. 4 mm long, 7 mm wide, shape unknown, bivalved; seeds not seen. Figs. 5A, 12F.

Phenology. Collected in flower in January, February, July, August, October, and November; in fruit only in October.

Distribution (Fig. 23). Ecuador (Esmeraldas, Pichincha); humid ravines and undisturbed low-elevation cloud forest; (100–) 500–700 m.

ADDITIONAL SPECIMENS EXAMINED. **Ecuador**. ESMERALDAS: Montañas de Mache, 35 km W of Quinindé, Bilsa Biological Station, *Bass & Pitman* 203 (US), *Clark & Adnepos 58* (MO, QCNE).—PICHINCHA: 2 km SE of Santo Domingo de Los Colorados, along Río Verde, Santa Marta Cooperative, *Dodson 7442* (MO, QCNE, SEL); Quito–Puerto Quito rd, Km 113, ENDESA Forest Reserve, *Jaramillo 7003* (AAU); Santo Domingo–Puerto Limón rd, Km 23, Colorado community "Congoma Grande," *Kvist 40701* (AAU).

The obovate and often nearly orbicular leaves with obtuse apices set *G. crispus* apart from all other species, except some populations of *G. calcaratus* subsp. *oncogastrus* also in western Ecuador. These latter plants usually differ by having inflorescences of more than three flowers, and by having smaller corollas (less than 2.5 cm long) (Fig. 6C). In addition, in *G. crispus* the stomatal clusters on the lower leaf surfaces are extraordinarily obvious. This is partly due to the villous indumentum that covers the lower leaf surfaces, except for the stomatal clusters. The indumentum appears to frame each stomatal cluster. The leaves of *G. crispus* are more or less appressed to the ground and almost form a rosette, the blades are mostly anisophyllous, and the upper leaf surfaces appear nearly black. These characters in combination with the urceolate, large, shiny, bright, yellow-orange to red corollas (Fig. 5A) make *G. crispus* an extremely conspicuous and beautiful species.

Gasteranthus crispus is only known from a few localities in the provinces of Pichincha and Esmeraldas of western Ecuador. It mainly occurs in low-elevation cloud forests between 500 and 700 m elevation, but has been found in humid ravines as low as 100 m elevation (Fig. 14). Currently it may survive only in the forests at the Bilsa Biological Station in Esmeraldas, where *G. crispus* is common in the shaded forests (pers. obs.).

11. Gasteranthus delphinioides (Seemann) Wiehler, Selbyana 1: 155. 1975. *Drymonia delphinioides* Seemann, Bot. Voy. Herald 186. 1854. *Besleria delphinioides* (Seemann) Leeuwenberg, Acta Bot. Neerl. 8: 49. 1959.—TYPE: COLOMBIA. Chocó: Cabo Corrientes, 1848, *Seemann 1054* (holotype: BM!; isotype: K!).

Besleria subcoriacea C. V. Morton, Contr. U.S. Natl. Herb. 29: 28. 1944.—TYPE: COLOMBIA. Chocó: Bahía Solano, dense forest along Quebrada Jellita, 50–100 m, 22 Feb 1939, *Killip & García 33486* (holotype: US!; isotypes: A! BM! COL! F! S!).

Besleria callista Standley & L. O. Williams, Ceiba 1: 248. 1951.—TYPE: COSTA RICA. Puntarenas: forested hills along the upper Río Piedras Blancas, vicinity of Río Esquinas, 30 m, 3 Aug 1950, *Allen 5589* (holotype: US!; isotypes: BH! CAS! DS! EAP-2 sheets! F! MO! US! WIS!).

Gasteranthus maculatus Wiehler, Selbyana 5: 87. 1978.—TYPE: Specimens from cultivated material, 16 Nov 1978, *Wiehler 78167* (holotype: SEL!; isotypes: K! MO! US!); plants brought into cultivation by Dressler from Panama, Cerro Pirre, Prov. Darién.

Herbs to subshrubs; stems erect, to 1 m tall, to 8 mm in diameter, distally mostly arachnoid to tomentose, less commonly villous or strigillose, basally glabrescent, cortex often appearing somewhat muricate, internodes 1–5 cm long, mostly terete, sometimes quadrangular. Leaves opposite, subisophyllous; blades 7–27 cm long, 3–12 cm wide, usually obovate to oblanceolate, occasionally elliptic, often decurrent, base attenuate to acute or cuneate, rarely obtuse, apex usually acute, margin serrulate, crenulate, or subentire, flat but usually somewhat muricate, adaxially green to dark green, glabrous, abaxially lighter green with conspicuous darker veins or reddish, usually more or less villous along the veins including the tertiary ones surrounding the stomatal clusters, lateral veins mostly 5–8 but occasionally to 12 per side, stomatal clusters conspicuous; petioles usually 0.5–2.5 cm long but occasionally to 4.5 cm, villous to glabrescent. Inflorescences open cymes of mostly 1–3 but occasionally with up to 12 flowers, pilose, sparsely pilose, or glabrous; peduncles 3–9 cm long, pedicels 1–7 cm long. Calyx light green, outside mostly pilose but occasionally villous, mostly with some glandular trichomes, sometimes nearly glabrous, inside apparently always more or less glandular-pilose, lobes free nearly to base, subequal to unequal, each of the lobes 7–20 mm long, 6–15 mm wide, broadly ovate with acute to obtuse apex, dorsal lobe keeled and the largest, ventral lobes the smallest, erose to crenulate, usually glandular to pilose, dorsal lobe directed backward and down below and around spur, ventral lobes directed forward and appressed to corolla, lateral lobes directed upward. Corolla 3.5–7.5 cm long, broadly funnelform to campanulate with large limb and long curved spur, tube not bent downward, 2.5–3.5 cm, widened from 4–9 mm at base to 18–38 mm in throat, white to yellow, outside apparently always more or less glandular-pilose, inside more or less glandular-puberulent, throat usually with a red-purple blushing or stripes inside, either from the three lower lobes or from all five lobes and continuing down through the inside of the corolla; spur 15–30 mm long, pointed, gradually narrowing from 4–8 mm to 1–2 mm wide, bent strongly downward and more or less strongly curved; throat 18–38 mm in diameter at widest point; limb varying from white-yellow to purple or red particularly centrally and basally at the lobes, lobes subequal-unequal, lower three lobes and upper two lobes more or less fused making the limb somewhat bilabiate, the upper (dorsal) lobes considerably smaller than the lower (ventral and lateral) lobes, rotund to acute, 4–10 mm long, 6–15 mm wide (variable and difficult to measure in dried material), glabrous. Filaments 15–21 mm long, adnate to base of corolla tube for 3–5 mm, nearly glabrous-puberulent; anthers ca. 2.5 mm long, 2 mm wide. Nectary dorsal, bilobed, ca. 2 mm high and directed down in spur, pubescent, more densely so distally. Ovary ca. 6 mm long, 5 mm in diameter, puberulent-velutinous; style 12–14 mm long, puberulent-velutinous; stigma stomatomorphic to weakly bilobed. Capsule ca. 5 mm long, 8 mm wide, flattened, splitting into 2 or 4 valves; seeds not seen. Figs. 3B, 11G.

Phenology. Collected in flower from January to December; in fruit during January, April, and October.

Distribution (Fig. 24). Colombia (Antioquia, Chocó, Nariño, Risaralda), Costa Rica (Puntarenas, San José), Panama (Bocas del Toro, Colón, Darién, Panamá, San Blas, Veraguas); wet slopes and stream banks in densely shaded wet forests; sea level to 2000 m.

REPRESENTATIVE SPECIMENS. **Costa Rica**. PUNTARENAS: above Golfito, *Dressler & Andrews 4171* (SEL, US).—SAN JOSÉ: ca. 1 mi beyond divide between San Isidro del General and coastal town of Dominical, *Croat 35287* (US). **Panama.** BOCAS DEL TORO: Río San Pedro, *Gordon 33C* (MO); Río San Pedrito, *Gordon 65C* (MO, US).—COLÓN: along Río Escandaloso near Mina Boquerón No. 2, *Antonio 1312* (US); lower Río Guanche, *Dressler 4220* (SEL, US).—DARIÉN: Cerro Tacarcuna massif between summit camp on westernmost ridge and Pico Tacarcuna, *Gentry et al. 16925 p.p.* (MO, US; the duplicate at COL is *G. wendlandianus*); Parque Nacional Darién, vic. Pirre station, *Herrera 678* (US).—PANAMÁ: El Llano–Cartí hwy, 19 km N of El Llano, *Dressler 4277* (F-2 sheets, NY, SEL-2 sheets, US).—SAN BLAS: Yar Bired (Cerro San José), continental divide between Cangandi and San José, *Nevers & Herrera 6904* (MO, US); Río Taindi, 6 km above confluence with Río Mandinga, *Nevers & Herrera 7669* (MO, US).—VERAGUAS: trail from Coquyito mine down to Río Barrera junction with Río Concepción, *Hammel 5219* (MO).

Colombia. ANTIOQUIA: Hoya del Río León or Bacubá, from Villa Arteaga to Chigorodó, Río Porrosa, La Pradera, *Cuatrecasas & Willard 26184* (COL, US); Mpio. Frontino, Correg. Nutibara, reg. Murrí, Nutibara–La Blanquita rd, *Luteyn et al. 12141* (HUA); Mpio. Mutatá, Correg. Longani, 2 km N of Mutatá, *Callejas et al. 5656* (HUA, US).—CHOCÓ: Tutunendo–El Carmen Hwy, Old Río Atrato, *Forero et al. 5880* (COL); Río Mutatá, tributary of Río El Valle between base of Alto de Buey and mouth of river, *Gentry & Fallen 17295* (COL, MO, US).—NARIÑO: CONIF Silvicultural station, "La Espriella," E of Tumaco, *León et al. 1269* (U, US).—RISARALDA: Mpio. Mistrató, correg. Puerto de Oro, Vereda Chirrincha, *Fernández Alonso et al. 9054* (COL); Mpio. Mistrató, San Antonio de Chami–Mistrató rd, Quebrada Sutu, *Fernández Alonso et al. 10168* (COL, HUA, US).

The large funnelform to campanulate corollas with a long, narrow, and downward-pointed spur (Fig. 3B) set *G. delphinioides* apart from all sympatric species. Only the two rare Ecuadorian species, *G. atratus* (Fig. 3A) and *G. tenellus* (Fig. 3C), have a similar

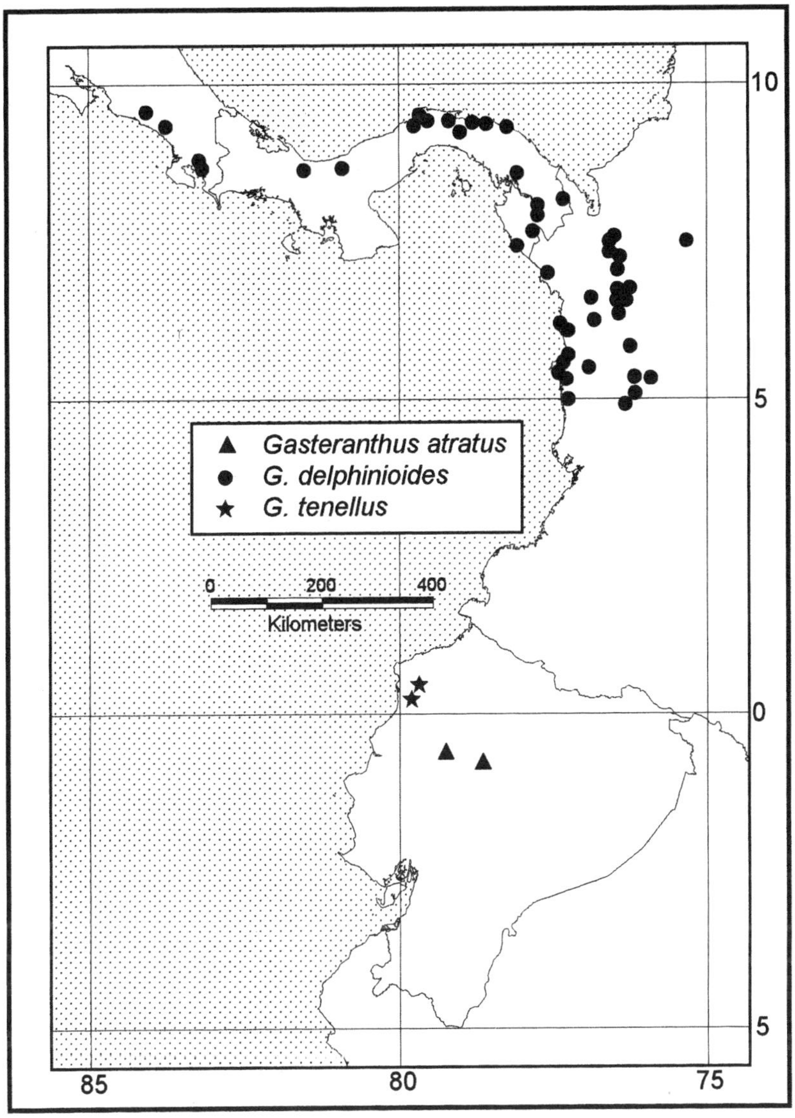

FIG. 24. Distribution of *Gasteranthus atratus*, *G. delphinioides*, and *G. tenellus*.

spur. The former differs by having strongly rugose leaves usually appressed to the ground, and the latter (Fig. 37) differs by having epedunculate inflorescences. Two species sympatric with *G. delphinioides*, *G. dressleri* in Panama (Fig. 3D) and *G. herbaceus* in Colombia (Fig. 3E), have equally large corollas with a wide throat and a broad limb. They both differ by having a much wider, blunt spur, by having the corolla tube bent sharply downward ca. 10 mm above the attachment of the calyx, and by the absence of conspicuous stomatal clusters. In addition, *G. dressleri* and some populations of *G. herbaceus* have shoots appressed to the ground, and particularly the latter species usually has longer petioles.

Gasteranthus delphinioides occurs mainly from Costa Rica in the north to the Colombian Pacific coast of the department of Chocó in the south. *Gasteranthus delphinioides* is particularly common in Panama and in the Colombian department of Chocó; in 1995 during field work along the coast of Chocó it was found along virtually all the streams in ravines in the forest (pers. obs.). The species is mainly found in the lowland with most collections made from below 500 m elevation, but it occurs at sites up to 1400 m elevation in the Colombian departments of Chocó and Antioquia and the Panamanian province of Darién; one anomalous, extremely small-flowered collection was obtained in Darién at ca. 1700 m. In Risaralda Department, Colombia, the species mainly occurs at elevations between 1500 and 2000 m.

12. Gasteranthus dressleri Wiehler, Selbyana 2: 127, pl. 35C. 1977.—TYPE: PANAMA. Colón: about 1–2 km upstream on Río Guanche from bridge and road to Portobelo, 10 Aug 1971, *Wiehler & Dressler 71165* (holotype: SEL!; isotype: US!).

Herbs; stems appressed to ground, not exceeding 10 cm tall, to 3 mm in diameter, pilose, internodes rather irregular, only a few mm long. Leaves opposite, subisophyllous; blades 5–17 cm long, 3–10 cm wide, ovate to elliptic, base obtuse to acute, apex acute to obtuse, margin serrate, more or less rugose, adaxially dark green, pilose, abaxially lighter green, villous along veins, lateral veins 5–7 per side, stomatal clusters not conspicuous; petioles 0.5–3 cm long, pilose. Inflorescences open cymes of 2–5 flowers, pilose; peduncles 0.5–1 cm long; pedicels 0.5–3 cm long. Calyx green, outside pilose to villous, inside glabrous, lobes free nearly to base, subequal, each of the lobes 13–17 mm long, 4–7 mm wide, lanceolate with acute apex, entire, dorsal lobe directed backward below spur, ventral and lateral lobes directed upward. Corolla 3.5–4.5 cm long, broadly funnelform to campanulate with large limb, strongly bent downward ventrally 5–10 mm from base, tube ca. 2 cm long straight distance from base to throat, widened from 5–7 mm at base to 20–25 mm, white, outside pilose, inside puberulent, marked in throat with red lines and a ventral yellow spot; spur 8–12 mm long, to 8 mm wide, apex blunt, directed downward; throat 20–25 mm in diameter; limb white, lobes 8–15 mm long, 8–12 mm wide, subequal to unequal, rotund, the ventral lobe largest, puberulent. Filaments 13–15 mm long, adnate to base of corolla tube for ca. 5 mm, glabrous; anthers ca. 2 mm long, 1.5 mm wide. Nectary dorsal, bilobed, ca. 1 mm high, glabrous. Ovary ca. 4 mm long, 2 mm in diameter, sparsely puberulent; style ca. 15 mm long, glabrous; stigma stomatomorphic. Capsule ca. 7 mm long, 8 mm wide, subglobose, bivalved; seeds not seen. Figs. 3D, 13I.

Phenology. Collected in flower in August, October, and November; in fruit in January, May, and October.

Distribution (Fig. 28). Panama (Colón); shaded streamsides and banks in lowland wet forests; 0–400 m.

ADDITIONAL SPECIMENS EXAMINED. **Panama**. COLÓN: Río Guanche, *Antonio 4777* (US), *Davidse & D'Arcy 10081* (MO, US), *Knapp 1421* (US), *Mori et al. 6466* (MO), *Skog et al. 4180* (MO, US), *Sytsma 1639* (US).

Gasteranthus dressleri is similar to *G. herbaceus* from Colombia. Both differ from all other species of *Gasteranthus* by having broadly funnelform to campanulate corollas that combine a large limb, ventrally bent sharply downward ca. 10 mm above the attachment to the calyx, with a wide, downward-bent, blunt spur (Fig. 3D). *Gasteranthus dressleri* differs from *G. herbaceus* (Fig. 3E) in having shortly pedunculate rather than epedunculate corollas, by always having congested leaves appressed to the ground, and by having more pilose, somewhat rugose upper leaf surfaces.

Gasteranthus dressleri is endemic to Panama and only known from a few localities near Río Guanche in the province of Colón.

13. Gasteranthus epedunculatus L. E. Skog & L. P. Kvist, sp. nov.—TYPE: COLOMBIA. Norte de Santander: Eastern Cordillera, Region of Sarare, Río Margua valley, Cabaceres del Río Negro, between El Amparo and La Mesa, 1400–1700 m, 7 Nov 1941, *Cuatrecasas 12866* (holotype: COL!; isotypes: F! US!).

Species insignis floribus epedunculatis corollisque ecalcaratis a speciebus cognitis bene distincta.

Herbs; stems decumbent with erect shoots, to 70 cm tall, to 6 mm in diameter, sparsely pilose to villous, internodes 1–2 cm long, terete. Leaves opposite, subisophyllous; blades 7–17 cm long, 3–6 cm wide, ovate to elliptic, base acute, apex acute to acuminate, margin serrate to subentire, flat, adaxially dark green, pilose or pubescent as juvenile and later glabrous, abaxially lighter green, occasionally with a violet sheen, pilose or glabrous, veins villous or tomentose, lateral veins 7–8 per side, stomatal clusters not conspicuous; petioles 1–4 cm long, sparsely pilose to villous. Inflorescences epedunculate cymes of 1–3 flowers, glabrescent, glandular-pilose or villous; peduncles absent; pedicels 2.5–5 cm long. Calyx green, outside glabrescent, puberulent or pilose particularly along a darker central longitudinal keel, inside sparsely pubescent to glabrous, lobes basally connate 1–2 mm, subequal, each of the lobes 15–25 mm long, 3–8 mm wide, lanceolate with acuminate apex, entire, dorsal, ventral, and lateral lobes all directed forward. Corolla 2.5–3.8 cm long, funnelform with a large limb, tube appearing bent downward or at least bent away from the axis, 1.7–2.5 cm long, widened from 3–5 mm at base to 6–9 mm in diameter at widest point, yellow, outside glabrous, inside glandular-puberulent in throat, otherwise puberulent only at base, marked apparently with lines; spur absent (but tube dorsally gibbous); throat 6–9 mm in diameter; limb yellow, lobes unequal, rotund, dorsal lobes 6–8 mm long, 5–7 mm wide, lateral and ventral lobes 8–11 mm long, 8–10 mm wide, glabrous. Filaments 12–21 mm long, adnate to base of corolla tube for 4–6 mm, glabrous above attachment to corolla tube; anthers ca. 2 mm long, 1.5 mm wide. Nectary dorsal, semiannular, to 1 mm high, glabrous. Ovary ca. 4 mm long, 2 mm in diameter, glabrous; style 12–17 mm long, glandular-puberulent; stigma stomatomorphic. Capsule ca. 7 mm long, 7 mm wide, globose, bivalved; seeds ca. 0.4 mm long, 0.2 mm wide, ellipsoid, papillate. Figs. 2A, 11D, E, 25.

Phenology. Collected in flower and fruit only during November.

Distribution (Fig. 15). Colombia (Norte de Santander); on riverbanks in wet forest; 600–1700 m.

ADDITIONAL SPECIMENS EXAMINED. **Colombia**. NORTE DE SANTANDER: Sarare region, Quebrada de la China (affluence to Río Cubugón), between Santa Librada and El Caraño, *Cuatrecasas 12989* (AAU, F, US), *Cuatrecasas* 13005 (F, US).

The combination of epedunculate inflorescences and corollas without spurs distinguishes *G. epedunculatus* from all other species. *Gasteranthus herbaceus* is the only other epedunculate species, but its corolla has a large, downward-bent spur (Fig. 3E).

FIG. 25. *Gasteranthus epedunculatus*. A. Erect apex of decumbent stem. B. Inset: section of abaxial leaf surface. C. Corolla. D. Corolla opened to show stamens. E. Calyx. F. Calyx, pistil, and disk. (Based on: A, B, *Cuatrecasas 13005*; C–F, *Cuatrecasas 12866*.)

Gasteranthus anomalus is the only other species without a spur (and therefore has all five calyx lobes directed forward), but it differs by having congested leaves appressed to the ground and by having a red, urceolate corolla (Fig. 10B). The most similar species are the two rare Ecuadorian endemics, *G. mutabilis* (Fig. 2C) and *G. orientandinus* (Fig. 2D), both of which have similar funnelform corollas with a large limb but with a conspicuous spur. Both also differ from *G. epedunculatus* in having conspicuous stomatal clusters on the lower leaf surfaces.

There is considerable variation among the three collections of *G. epedunculatus*; particularly *Cuatrecasas 13005* differs by being much more pubescent, and having a shorter corolla tube and wider calyx lobes than the others.

Gasteranthus epedunculatus has been collected only locally on the eastern Andean slopes in the Colombian department of Norte de Santander near the border with Venezuela, considerably farther to the northeast than any other species of *Gasteranthus* (Fig. 1). Few botanists have visited this area since José Cuatrecasas obtained the three collections of *G. epedunculatus* in 1941.

14. Gasteranthus extinctus L. E. Skog & L. P. Kvist, sp. nov.—TYPE: ECUADOR. Los Ríos or Pichincha: Montañas de Ila, cloud forest along ridge line near La Centinela, Km 12, on road from Patricia Pilar to Flor de Mayo, 600 m, 4 Oct 1981, *Dodson 11595* (holotype: QCNE!; isotypes: MO! SEL!).

Species a Gasterantho crispo (Mansfeld) Wiehler foliis ellipticis, caulibus conspicue quidem apice villosis distincta.

Herbs; stems decumbent to erect, rising to 50 cm tall, to 3 mm in diameter, pilose to villous distally, internodes 0.5–1.5 cm long, terete to quadrangular. Leaves opposite, slightly anisophyllous; blades 5–11 cm long, 2–4.5 cm wide, elliptic, base cuneate to acute, apex acute, margin serrate and pilose, flat, adaxially green, sparsely appressed pilose, abaxially lighter green, pilose to villous, particularly on the veins, lateral veins 7–9 per side, stomata in conspicuously lighter green clusters; petioles 0.5–3 cm long, pilose-villous. Inflorescences open cymes of 2–4 flowers (rarely reduced to a single flower), pilose to villous; peduncles 3–4 cm long; pedicels 0.3–1.3 cm long. Calyx green, outside villous, inside sericeous, lobes basally connate ca. 1 mm, subequal, each of the lobes lanceolate to more or less ovate with acute to acuminate apex, dorsal lobe keeled, 10–12 mm long, 3–5 mm wide, erose, dorsal lobe directed backward surrounding spur, ventral lobes directed forward, lateral lobes directed forward or upward. Corolla 2.8–4 cm long, urceolate with a small limb and a ventral pouch, tube (defined as distance from base of corolla to throat) 1.5–2.5 cm long but distance to tip of pouch 2.5–3.5 cm and thus protruding beyond throat, widened from 2–4 mm at base to maximally 0.8–1.8 cm (with pouch) in diameter at widest point, orange, outside puberulent, inside puberulent, particularly within pouch (ventrally), markings none; spur 3–5 mm long, blunt, straight; throat 3–6 mm in diameter; limb orange, lobes 1–2 mm long, 2–4 mm wide, subequal-unequal, acute-rotund, glabrous. Filaments 9–17 mm long, adnate to base of corolla tube for 3–7 mm, glabrous but glandular; anthers ca. 1.5 mm long, 1.5 mm wide. Nectary dorsal, broadly but shallowly bilobed, to 1 mm high, glandular-tomentose. Ovary ca. 5 mm long, 5 mm in diameter, velutinous; style ca. 10 mm long, sparsely puberulent; stigma stomatomorphic to bilobed. Capsule and seeds not seen. Figs. 5B, 26.

Phenology. Collected in flower in July, August, and October.

FIG. 26. *Gasteranthus extinctus*. A. Habit. B. Inset: section of abaxial leaf surface. C. Corolla. D. Corolla opened to show stamens. E. Calyx. F. Calyx, pistil, and disk. (Based on: A, *Dodson & Dodson 15867*; B, *Dodson 11595*; C–F, *Dodson et al. 7117*.)

Distribution (Fig. 23). Ecuador (Los Ríos/Pichincha border area); in deep shade in cloud forest; 600 m.

ADDITIONAL SPECIMENS EXAMINED. **Ecuador**. LOS RÍOS or PICHINCHA: Montañas de Ila, La Centinela, *Dodson & Dodson 6809* (SEL), *15867* (US), *Dodson et al. 7117* (MO, SEL),

The following combination of characters distinguishes *G. extinctus*: 1) urceolate, relatively large corollas (2.8–4 cm long) (Figs. 5B, 26C); 2) inflorescences with relatively short peduncles (maximally 4 cm long); 3) few flowers (2–4) per cyme; 4) a conspicuous pilose-villous indumentum on stems, inflorescences, and calyces; and 5) fairly small, elliptic leaves (maximally 11 cm long). The most similar species may be the widespread *G. calcaratus*. The sympatric *G. calcaratus* subsp. *oncogastrus*, however, always has smaller flowers than *G. extinctus*, but the Colombian *G. calcaratus* subsp. *calcaratus* occasionally has equally large flowers (Fig. 6A). *Gasteranthus calcaratus*, however, never has a conspicuous villous indumentum, and usually has inflorescences with longer peduncles and more flowers than *G. extinctus*. Vegetatively the most similar species may be the equally rare *G. mutabilis* (Fig. 2C), known only from the montane forests of the Ecuadorian Pichincha province, but this species has funnelform rather than urceolate corollas.

Gasteranthus extinctus is known only from a single range of western Andean foothills in Los Ríos (or Pichincha) in western Ecuador. All four collections come from a 600 m tall ridge that was once covered with low-elevation cloud forest; the forest has been totally cleared, likely causing the extinction of this species.

15. Gasteranthus glaber L. E. Skog & L. P. Kvist, sp. nov.—TYPE: ECUADOR. Pichincha: old Quito–Santo Domingo road (between new road and Chiriboga), on slopes in wet ravines, 1900 m, 28 Oct 1995, *Mendoza-T. et al. 624* (holotype: QCNE!; isotypes: AAU! B! BM! CAS! COL! E! F! GB! GH! HUA! LE! MO! NY! P! PE! QCA! S! SEL! TULV! US! VEN! WU!).

Plantae glabrae inflorescentiis calycibusque rubro-purpureis corollis parvis 1.5–2.2 cm longis ventricosis vel urceolatis in calycibus occultis praeter lobo subatratis a speciebus cognitis bene distincte.

Herbs to subshrubs; stems erect, rising to 1.2 m tall, to 7 mm in diameter, glabrous, internodes 2–8 cm long, fairly smooth, not wrinkled, quadrangular. Leaves opposite, isophyllous; blades 7–19 cm long, 4–10 cm wide, ovate to elliptic, rarely narrowly elliptic to oblanceolate, base cuneate to acute, apex acute to acuminate, margin serrate, flat, adaxially dark green, glabrous, abaxially lighter green with conspicuously darker veins, glabrous, lateral veins 6–9 per side, in addition intersecondary veins particularly conspicuous, stomatal clusters not conspicuous; petioles 1–3 cm long, glabrous. Inflorescences congested cymes of 3–15 flowers, almost perpendicular to the stem, glabrous, usually wine-red; peduncles 5–10 cm long; pedicels 0.2–1.2 cm long. Calyx red, rarely yellow, glabrous, lobes basally connate 1–2 mm, subequal, each of the lobes ovate to semiorbicular, apex acute to obtuse or rotund, dorsal lobe keeled, 10–16 mm long, 8–12 mm wide, mostly entire, occasionally serrulate, dorsal lobe directed backward with spur, ventral lobes directed forward and appressed to corolla, lateral lobes directed upward. Corolla 1.5–2.2 cm long, urceolate, sometimes with a relatively small limb and a ventral pouch, sometimes approaching ventricose, tube 1.3–1.8 cm long, widened from 2–4 mm at base to maximally 0.7–0.9 cm in diameter at widest point, dark red, outside glabrous, inside glabrous, markings none (except for the very dark limb); spur 1–4 mm long, blunt, straight; throat 3–5 mm in diameter; limb blackish, lobes ca. 1.5 mm long, ca. 2 mm wide, subequal, acute-rotund, glabrous. Filaments 8–10 mm long, adnate to base of corolla tube for 1–3 mm, glabrous; anthers ca. 1.5 mm long, 1 mm wide. Nectary dorsal, semiannular, to 1 mm high but sometimes much reduced, sparsely pubescent. Ovary ca. 4 mm long, 3 mm in diameter, glabrous; style 6–9 mm long, glabrous; stigma stomatomorphic to weakly bilobed. Capsule ca. 8 mm long, 1.2 mm wide, flattened, dehiscence pattern unknown; seeds ca. 0.3 mm long, 0.2 mm wide, elliptic, somewhat papillate. Figs. 7C, 27.

Phenology. Collected in flower from January to April, June to December; in fruit in April and October.

Distribution (Fig. 19). Colombia (Nariño) and Ecuador (Carchi, Imbabura, Pichincha); on slopes in wet ravines in shady mature montane forests; (1400–) 1800–2100 (–2700) m.

ADDITIONAL SPECIMENS EXAMINED. **Colombia**. NARIÑO: Mpio. de Pasto, Daza, *Benavides 2884* (MO); Mpio. Ricaurte, 7 km from Chucunés, Natural reserve La Planada, *Benavides 10225* (MO), *10914* (US); La Planada, Salazar Finca 7 km above Ricaurte, *Gentry et al. 35147* (COL, MO, US); La Planada Reserve, 7 km from Chucunés, *Gentry & Keating 59693* (MO, US), *Gentry et. al. 30569* (MO-2 sheets, US), *Giraldo 11* (HUA), *36* (HUA), *74* (HUA); La Planada, *Restrepo 499* (US), *541* (US), *Stiles 575* (COL-2 sheets). **Ecuador**. CARCHI: Espejo, between Las Juntas and La Cabaña del Corazón, *Clark et al. 2423* (QCNE, US), *Clark & Dunn 2462* (AAU, COL, E, MO, QCNE, SRP, US, VEN), *2467* (QCNE, US); Tulcán–Maldonado rd, 13 km SE of Maldonado, *Harling & Andersson 12349* (GB, SEL); ridge NE of Quindi's finca, *Hoover 2020* (MO, QCA); along Tulcán–Maldonado rd, *Luer et al. 2673* (SEL); above Maldonado, *Luer et al. 3365* (SEL); 62–75 km W of Tulcán, *Luteyn & Cotton 10873* (US); rd to Chical, N of Carmen, *Palacios et al. 9735* (US); Espejo, Los Juntas–El Corazón, *Palacios 12422* (QCNE); Espejo, El Gualtal, faldas de Cerro Golandrinas Hembra, *Palacios & Clark 12623* (US); Maldonado–Tulcán rd, Km 20, *Werling & Leth-Nissen 359* (QCA).—IMBABURA: Cotocachi, Hda. La Florida, *Alvarez & Castro 660* (US).—PICHINCHA: Saloya, *Acosta Solís 5811* (F); old Quito–Santo Domingo rd, Chiriboga, Forest Reserve "La Favorita," *Cerón et al. 7912* (GB, US); old Quito–Santo Domingo rd, Km 63, *Dodson & Dodson 11834* (MO, QCNE, SEL, US); Mindo, Nono–San Miguel de los Bancos, *Hirtz 2159* (MO); between Chiriboga and Empalme, Km 59, *Jaramillo 5965* (QCA [mixed with *Gasteranthus mutabilis* and several other duplicates of that species have the same number], US); Floristic-Ecological Reserve "Río Guajalito," *Jaramillo 7773* (QCA), *Jaramillo & Zak 577* (AAU, NY), *7773* (QCA); Mindo, *Luer et al. 4735* (SEL); old Quito–Santo Domingo rd, Las Palmeras, *Luteyn & Berg 14360* (NY, QCA, US); Floristic-Ecological Reserve "Río Guajalito," *Quintana 22* (QCA), *Zak & Jaramillo 577* (AAU, COL, MO, QCA, US).

Gasteranthus glaber is totally glabrous (except for the sparsely pubescent nectary). It has more or less red or purple inflorescences and calyces, and small (1.5–2.2 cm long) ventricose to urceolate corollas nearly hidden by the surrounding calyces (Fig. 7C), except for the usually dark to nearly black limbs. The species is also characterized by usually having conspicuous darker veins on the lower leaf surfaces and particularly well-developed intersecondary veins, as well as by having the peduncles nearly perpendicular with the stems. This contrasts with the widely distributed *G. corallinus* (Fig. 21), which has similar-shaped small corollas, but which are arranged in nearly vertical long-pedunculate inflorescences. *Gasteranthus corallinus* also differs from *G. glaber* by having stems with a conspicuous indumentum, and by having relatively uniformly red or yellow corollas. *Gasteranthus carinatus* from western Ecuador and *G. imbricans* from Panama and Costa Rica also share similarities with *G. glaber*. *Gasteranthus carinatus* is the only other species in the genus that is nearly glabrous, and *G. imbricans* initially appears glabrous although it does have an inconspicuous indumentum along the veins of the lower leaf surfaces. The two species both have equally small corollas (1.5–2.5 cm long), but these are subventricose to funnelform rather than urceolate (Fig. 8A, C). *Gasteranthus carinatus*

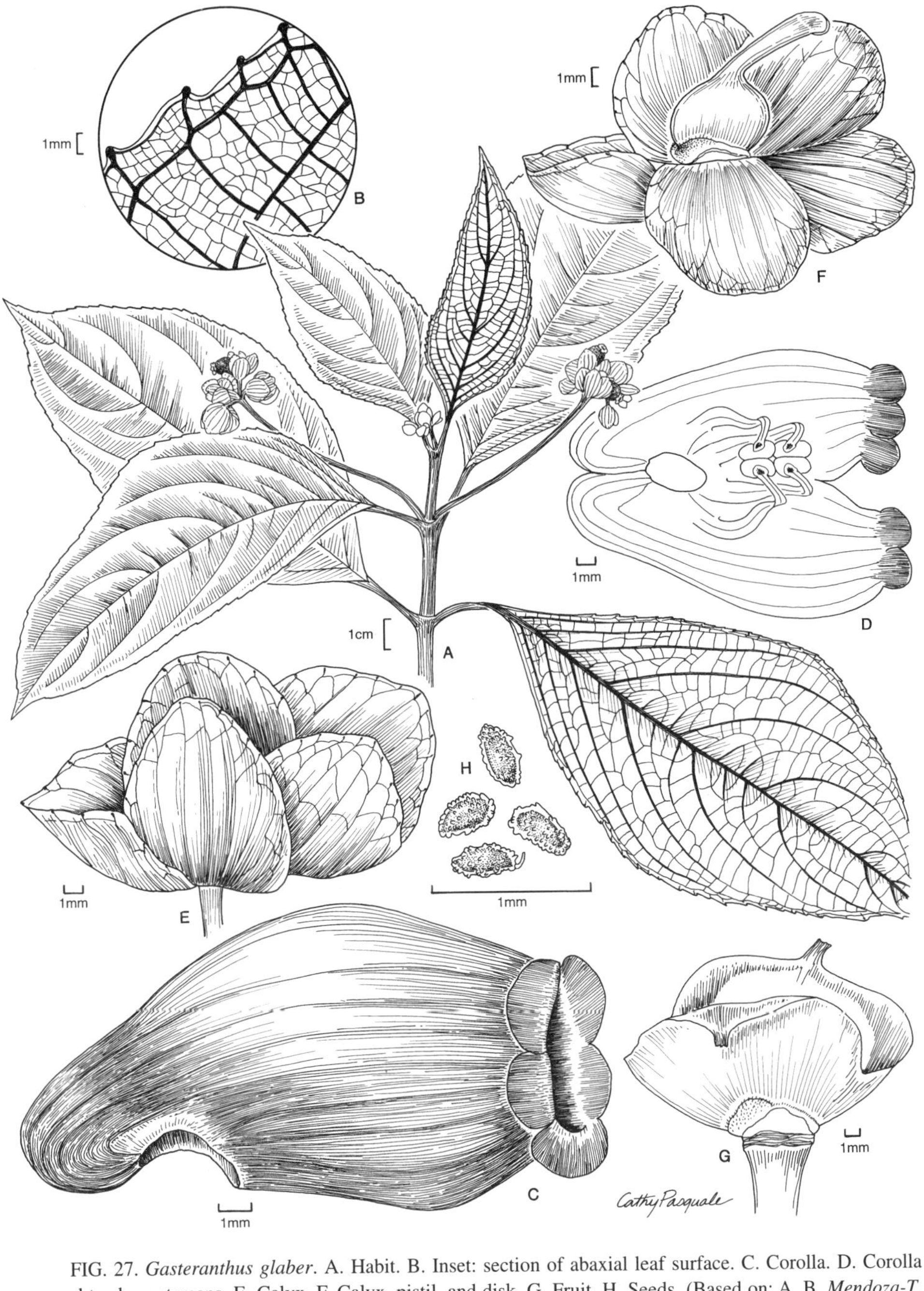

FIG. 27. *Gasteranthus glaber*. A. Habit. B. Inset: section of abaxial leaf surface. C. Corolla. D. Corolla opened to show stamens. E. Calyx. F. Calyx, pistil, and disk. G. Fruit. H. Seeds. (Based on: A, B, *Mendoza-T. et al. 624;* C–F, *Dodson & Dodson 11834*; G, H, *Clark & Dunn 2462*.)

and *G. imbricans* are also both shrubbier than *G. glaber*, and have stems with a wrinkled bark rather than a smooth surface. The rare *G. otongensis*, known only from its type locality in Cotopaxi province south of the range of *G. glaber*, also has small corollas nearly hidden by the more conspicuous red calyx lobes; however, nearly all its surfaces have a vestiture.

Gasteranthus glaber occurs in the western Andean montane forests from the Ecuadorian province of Pichincha in the south to the Colombian department of Nariño in the north, and it is these remote populations that are most similar. The intervening populations from the Ecuadorian province of Carchi differ in their more open inflorescences with fewer flowers (mostly only 2–3), the more or less serrulate rather than entire calyx lobes, and the sometimes narrowly elliptic rather than elliptic leaves.

16. Gasteranthus herbaceus (C. V. Morton) Wiehler, Selbyana 1: 155. 1975. *Besleria herbacea* C. V. Morton, Contr. U.S. Natl. Herb. 26: 464. 1939.—TYPE: COLOMBIA. Cauca: El Tambo, 800 m, 3 Jul 1936, *Sneidern 762* (holotype: S!).

Herbs; stems decumbent to erect or appressed to ground, to 50 cm tall, to 7 mm in diameter, glabrous to pilose or rarely villous distally, internodes nearly absent to 12 cm long, mostly quadrangular. Leaves opposite, more or less anisophyllous; blades 4–27 cm long, 2–15 cm wide, ovate to elliptic or obovate, base acute to obtuse, apex acute, margin serrate, rarely subentire, flat, adaxially green, appressed pilose to glabrous, abaxially lighter green, sparsely pilose to pilose or tomentose along veins, lateral veins 5–7 per side, stomatal clusters not conspicuous; petioles 2–11 cm long, pilose to villous. Inflorescences epedunculate cymes of mostly single flowers but occasionally up to 4 in same leaf axil, pilose to villous; peduncles absent; pedicels 2–7 cm long. Calyx green, outside pilose to villous particularly basally, inside glabrous, lobes free nearly to base, subequal, each of the lobes usually lanceolate, occasionally approaching ovate, apex acuminate, dorsal lobe keeled, 11–18 mm long, 4–9 mm wide, entire, dorsal lobe directed backward below spur, ventral lobes directed forward and appressed to corolla, lateral lobes directed upward. Corolla 3.5–5.5 cm long, broadly funnelform to campanulate with large limb, strongly bent downward ventrally ca. 5 mm from base, tube 1.8–2.5 cm long, diameter widened from 5–10 mm at base to 1.5–2.5 cm in throat, white, occasionally pink or red, outside pilose particularly basally, inside puberulent in throat, markings in throat yellowish violet with violet-purple lines; spur 6–12 mm long, 5–10 mm wide, blunt, directed downward; throat 1.5–2.5 mm in diameter; limb white, rarely red, lobes 6–15 mm long, 8–14 mm wide, subequal-unequal, rotund, the ventral and lateral lobes often the largest, puberulent to glabrous along edge. Filaments 14–21 mm long, adnate to base of corolla tube for 3–7 mm, glabrous; anthers ca. 2 mm long, 1.5 mm wide. Nectary dorsal, bilobed, to 1.5 mm high, sparsely villous. Ovary 4–5 mm long, 3–4 mm in diameter, puberulent; style 15–18 mm long, puberulent basally to glabrous distally; stigma stomatomorphic to weakly bilobed. Capsule ca. 5–7 mm long, 10 mm wide, flattened, bivalved, secondarily splitting into 4 valves; seeds ca. 0.2–0.3 mm long, 0.2 mm wide, elliptic, papillate. Figs. 3E, 13A, B.

Phenology. Collected in flower in February, March, May, July, August, and October to December; in fruit in February and May.

Distribution (Fig. 28). Colombia (Antioquia, Cauca, Chocó, Risaralda, Valle del Cauca); shaded streamsides or often on steep wet banks in dense forest; (100–) 1000 (–2000) m.

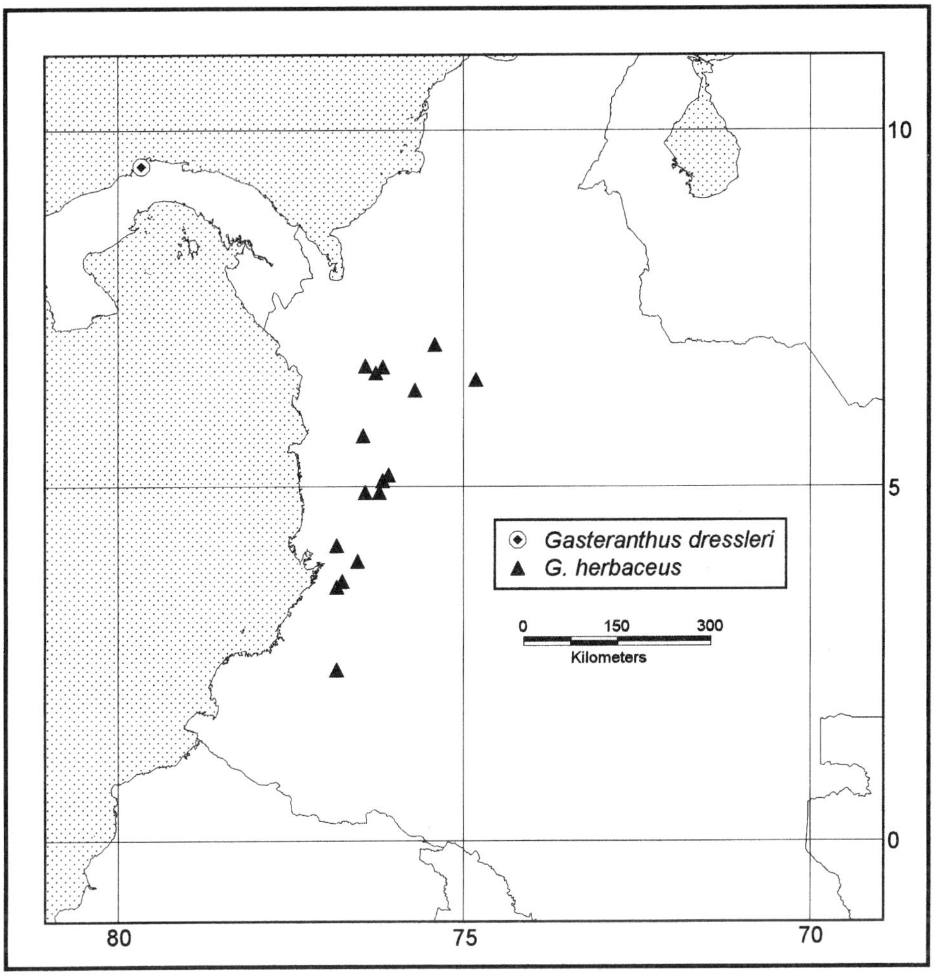

FIG. 28. Distribution of *Gasteranthus dressleri* and *G. herbaceus.*

ADDITIONAL SPECIMENS EXAMINED. **Colombia**. ANTIOQUIA: National Park "Las Orquideas," sector Tres Bocas, *Cogollo et al. 4043* (MO); Alto San Fermin, near Yarumal, *Core 615* (US); near Palmitas, W of Boquerón Pass, *Hatheway 1559* (BH); San José, *Kalbreyer 1511* (K); Alto Boquerón, *Kalbreyer 1511a* (K); Río Verde, *Kalbreyer 1787* (K); Mpio. Frontino, Murrí region, rd between Nutibara and La Blanquita, *Macdougal et al. 3899* (US); Mpio. Frontino, El Plateado near gold mine, *Luer et al. 9049* (SEL).—CAUCA: Dagua valley, between San José and El Roble, *Pittier 600* (US); El Tambo, *Sneidern 478* (US), *793* (NY, S).—CHOCÓ: Between Río Surama and Río Fiadó, *Forero et al. 3354* (MO); Mpio. San José de Palmar, Río Torito valley (affluent to Río Hábita), Santa Fé canyon, *Forero et al. 7306* (COL, MO, US); 11 km E of Tutunendo on Quibdo–Medellín rd, *Gentry & Fallen 17563* (COL).—RISARALDA: Mpio. Pueblo Rico, vereda La Selva, Río Taiba valley, Parque Nacional Natural Tatamá, *Escheverri et al. 511* (HUA); Pueblo Rico, *Sneidern 5464* (F, US-2 sheets).—VALLE DEL CAUCA: Río Dagua valley, San Juan canyon, below Queremal, *Cuatrecasas 22747* (F, US); left bank of Río Sanquininí, La Laguna, *Cuatrecasas 15579* (F); Alta Flor, Calima gorge, *Hugh-Jones 397* (K); old Cali–Buenaventura rd, Km 51, *Londoño & Kvist 181* (AAU, COL, TULV, US); Río Bravo, NW of Darién, *Robinson 85* (COL, K, US).

Gasteranthus herbaceus is conspicuous in its large funnelform-campanulate whitish or yellow corollas (Fig. 3E). The most similar species is *G. dressleri* from Panama, which differs mainly by having shortly pedunculate rather than epedunculate inflorescences (see *G. dressleri*, no. 12). Only one sympatric species, *G. delphinioides*, may have equally large and wide corollas (Fig. 3B), but this species has long-pedunculate inflorescences and corollas with a long and narrow curved spur, in contrast to the wider, blunt spur in *G. herbaceus*. The two species can also be distinguished in sterile condition; *G. herbaceus* has longer petioles than *G. delphinioides*, usually exceeding 2 cm, usually has a conspicuous pilose indumentum, and may have congested leaves appressed to the ground (as in *G. dressleri*).

Only two species of *Gasteranthus*, *G. herbaceus* and *G. epedunculatus* (from the Colombian department of Norte de Santander), have epedunculate inflorescences, but the latter species (Fig. 25) differs among other features from *G. herbaceus* by having corollas without a spur (see *G. epedunculatus*, no. 13).

Gasteranthus herbaceus is endemic to Colombia, where it occurs mainly on the western slopes of the Western Cordillera but also in Antioquia.

17. Gasteranthus imbaburensis M. Freiberg, Phyton (Horn) 36: 307, fig. 1B. 1996.—TYPE: ECUADOR. Imbabura: Los Cedros Biological Reserve, 1700 m, 4 Jul 1996, *Freiberg 96238* (holotype: QCA, photo: US!; isotypes: QCNE, ULM!).

Subshrubs; stems erect, to 2.5 m tall, to 6 mm in diameter, appressed pubescent, internodes 2–4 cm long, quadrangular. Leaves opposite, isophyllous; blades 15–20 cm long, 4–7 cm wide, ovate to elliptic, base acute, apex acute, margin serrulate, flat, adaxially dark green, sparsely strigose, abaxially green, densely appressed pilose on the veins, lateral veins 7–9 per side, stomatal clusters not conspicuous; petioles 2–4.5 cm long, appressed-pilose. Inflorescences open cymes of 6–15 flowers, strigillose; peduncles 10–20 cm long; pedicels 0.5–2 cm long. Calyx green, sparsely glandular-pilose outside, lobes basally connate ca. 2 mm, subequal, each of the lobes broadly ovate, apex acute, dorsal lobe ca. 11 mm long, 7 mm wide, ventral lobes ca. 12 mm long, 7 mm wide, lateral lobes ca. 10 mm long, 6 mm wide, serrate, dorsal lobe curved downward surrounding spur, ventral lobes directed forward and appressed to corolla, lateral lobes directed forward. Corolla ca. 2.6 cm long, urceolate with a small limb and a small ventral pouch not surpassing the throat, tube 1–1.5 cm long, widened from 4 mm at base to ca. 8 mm in diameter at widest point, vermilion, outside glandular-pilose, inside glabrous near throat, markings not seen; spur 7–8 mm long, blunt, straight; throat 3–6 mm in diameter; limb red, lobes ca. 2 mm long, 5 mm wide, subequal, rotund, glabrous. Filaments ca. 12 mm long, adnate to base of corolla tube for ca. 4 mm, puberulent; anthers ca. 3.5 mm long, 2 mm wide. Nectary dorsal, semi-annular, ca. 1 mm high, glabrous. Ovary ca. 8 mm long, 6 mm in diameter, glabrous; style ca. 9 mm long, glabrous; stigma weakly bilobed. Capsule and seeds not seen. Fig. 7B.

Phenology. Collected in flower during July.

Distribution (Fig. 20). Ecuador (Imbabura); shrub layer in primary premontane wet forest; 1700 m.

Gasteranthus imbaburensis is known only from the type collection. It resembles *G. adenocalyx*, from Antioquia in Colombia, in having glandular pilose calyx lobes (Fig. 2B); however, *G. imbaburensis* differs in having a pubescent stem rather than glabrous one, 7–9 veins per side on the leaves, larger calyx lobes, an urceolate reddish corolla (Fig. 7B), and a semiannular nectary, among other characters.

Gasteranthus imbaburensis is particularly similar to *G. timidus*. They both have conspicuously quadrangular nodes and internodes, and similar leaves. Both species also have flowers with a glandular indumentum, although particularly the calyces of *G. timidus* (Fig. 8D) have a much sparser indumentum than those of *G. imbaburensis*. In addition, the calyx lobes of *G. timidus* are much smaller, never exceeding the length of 4 mm, whereas those of *G. imbaburensis* are at least 5 mm long and usually exceed 8 mm, and tend to hide partly the corollas. The two species have separate ranges as well. *Gasteranthus imbaburensis* occurs in montane cloud forest, whereas *G. timidus* has a more southern and eastern distribution, mostly in low-elevation cloud forest, between 100 and 700 m elevation. *Gasteranthus carinatus* and *G. imbricans* may also have inflorescences and corollas somewhat similar to those of *G. imbaburensis*, but both these species are nearly to completely glabrous.

18. Gasteranthus imbricans (J. D. Smith) Wiehler, Selbyana 1: 155. 1975. *Besleria imbricans* J. D. Smith, Bot. Gaz. 25: 155. 1898.—TYPE: COSTA RICA. Limón: Forests of Shirores, Talamanca, 100 m, Feb 1895, *Tonduz 9192* (holotype: US!).

Besleria imbricans var. *uncinata* C. V. Morton, Publ. Field Mus. Nat. Hist., Bot. Ser. 18: 1152. 1938.—TYPE: COSTA RICA. Cartago: Las Vueltas, Tucurrique, 900–1000 m, Mar 1899, *Tonduz 13038* (holotype: US!; isotypes: BM! CR! P!).

Besleria panamensis C. V. Morton, Contr. U.S. Natl. Herb. 26: 460. 1939. *Gasteranthus panamensis* (C.V. Morton) Wiehler, Selbyana 1: 155. 1975.—TYPE: PANAMÁ. Bocas del Toro, 22 Apr 1921, *Carleton 205* (holotype: US!).

Subshrubs or shrubs; stems erect, to 2 m tall, to 8 mm in diameter, glabrous, internodes 1–8 cm long, terete to quadrangular, rather irregular and the cortex often strongly wrinkled. Leaves opposite, subisophyllous; blades 10–27 cm long, 5–16 cm wide, mostly elliptic to ovate or obovate, base cuneate to obtuse, often oblique, apex acute, margin subentire to irregularly serrate or dentate, flat, adaxially dark green, glabrous, abaxially lighter green, puberulent to pilose along veins, lateral veins 11–13 per side, stomatal clusters not conspicuous; petioles 1–6 (–11) cm long, puberulent to glabrous. Inflorescences congested cymes of 3–15 flowers, vertical, green, glabrous; peduncles 7–17 cm long; pedicels 0.5–2.0 cm long. Calyx orange, outside glabrous, inside with small brown glandules, lobes basally connate 1–4 mm, unequal and conspicuously imbricate, each of the lobes broadly ovate with obtuse apex, dorsal lobe and usually one of the ventral lobes keeled and the calyx thus asymmetric, 7–12 mm long, 7–11 mm wide, entire to sub-erose, dorsal lobe directed backward below spur and upward, ventral lobes more or less directed forward surrounding corolla, lateral lobes more or less directed forward surrounding corolla. Corolla 1.5–2.2 cm long, ventricose to funnelform with a small limb, nearly hidden by surrounding calyx, tube 0.9–1.5 cm long, widened from 4–6 mm at base to maximally ca. 10 mm in diameter at widest point, orange, outside glabrous, inside glandular-hairy mainly in throat, markings not seen; spur 4–6 mm long, to 3 mm wide, blunt, directed upward; throat 7–9 mm in diameter; limb orange, lobes 2–3 mm long, 2–4 mm wide, subequal, acute-rotund, glabrous. Filaments 8–11 mm long, adnate to base of corolla tube for 2–3 mm, glabrous; anthers ca. 2.5 mm long, 2.0 mm wide. Nectary dorsal, semi-annular but dorsally bilobed, to 1 mm high, villous distally, glabrescent basally. Ovary ca. 4 mm long, 2 mm in diameter, glabrous; style 6–9 mm long, glabrous; stigma weakly bilobed. Capsule ca. 5 mm long, 8 mm wide, broadly oblate, bivalved; seeds ca. 0.5 mm long, 0.3 mm wide, oblong, striate and papillate. Fig. 8C.

Phenology. Collected in flower from April to June, and in September, October, and December; in fruit during June and September.

Distribution (Fig. 20). Costa Rica (Cartago, Limón) and Panama (Bocas del Toro); along streamsides in shade in primary and secondary wet forests; 0–500 (–1000) m.

ADDITIONAL SPECIMENS EXAMINED. **Costa Rica**. CARTAGO: Río Pucuare, near Platanillo, *Williams 19515* (US).—LIMÓN: Río Sixaola drainage, 1–3 km N of Bribri, *Antonio 746* (F); Bribri, *Burger & Antonio 10964* (F), *10964A* (F); 7 km SW of Bribri, *Gómez et al. 20321* (MO, US); Río Segundo, Asunción, *Gómez & Herrera 23491* (MO, US); Río Peje, Asunción, *Gómez & Herrera 23515* (MO, US); Río Sandbox, Talamanca, *Gómez-Laurito 10463* (F), *Gómez-Laurito & Vargas 11960* (MO); Cantón de Talamanca, Amubri, right bank of Río Lari, *Herrera 3048* (F, US). **Panama**. BOCAS DEL TORO: 15 km S of the town Changuinola, *Antonio 3115* (MO, US); Sursuba, Río Changuinola, *Dwyer 2899* (MO), *4387A* (MO), *4900* (MO); Quebrada Hurón, *Kirkbride & Duke 450* (MO); Río Teribe, near Quebrada Lukulon, *Kirkbride & Duke 508* (MO-2 sheets, NY), *Kirkbride & Duke 516* (MO); Duwebdulup Peak, N of Río Teribe, *Kirkbride & Duke 573* (MO).

The yellowish, glabrous, broad, imbricate calyx lobes, which nearly enclose the fairly small (1.5–2.2 cm long) ventricose corollas (Fig. 8C), easily distinguish *G. imbricans*. In addition, the species is rather shrubby, and has a wrinkled, glabrous bark and large leaves with many secondary veins (11–13 per side). The spur of the corolla is directed upward, a feature that is otherwise only found in *G. recurvatus*, but this latter species has larger (3–3.5 cm long) urceolate corollas (Figs. 10C, 36).

Morton in his 1939 revision of *Besleria* included the unpublished name "Besleria imbricans var. arguta" in the key and in the index. This name is an error for *Besleria imbricans* var. *uncinata* C. V. Morton, which he had published the previous year (Morton 1938). Presumably, he had the two manuscripts in preparation at the same time, and neglected to correct the name in his 1939 revision.

19. Gasteranthus lateralis (C. V. Morton) Wiehler, Selbyana 1: 155. 1975. *Besleria lateralis* C. V. Morton, Contr. U.S. Natl. Herb. 26: 469. 1939.—TYPE: ECUADOR. Pichincha: Andes of Quito, 1855, *Couthouy s.n.* (holotype: GH!; isotype: GH!).

Besleria venusta C. V. Morton, Contr. U.S. Natl. Herb. 26: 468. 1939. *Gasteranthus venustus* (C. V. Morton) Wiehler, Selbyana 1: 155. 1975.—TYPE: ECUADOR. Pichincha: Tandagú, 1500 m, Jul 1920, *Heilborn 757* (holotype: S!).

Gasteranthus giganteus M. Freiberg, Phyton (Horn) 36: 304, fig. 1A. 1996.—TYPE: ECUADOR. Imbabura: Los Cedros Biological Reserve, 1400–1700 m, 22 Mar 1996, *Freiberg 96022* (holotype: QCA!, photo: US!; isotypes: QCNE, ULM!).

Herbs to subshrubs; stems erect, to 1.5 m tall, conspicuously thicker at nodes than between, to 1 cm in diameter, glabrescent to arachnoid or tomentose distally, internodes 2–8 cm long, quadrangular. Leaves opposite, isophyllous; blades 11–26 cm long, 5–13 cm wide, obovate to oblanceolate, base more or less attenuate to acute, apex acute to obtuse, margin serrate, flat, adaxially green, glabrous, abaxially lighter green, arachnoid to tomentose at the veins, lateral veins 6–10 per side, stomatal clusters not conspicuous; petioles up to 1.5 cm but rarely exceeding 0.5 cm long, arachnoid-tomentose. Inflorescences relatively congested cymes of 2–10 flowers, pilose to villous or arachnoid; peduncles 7–19 cm long; pedicels mostly ca. 0.5 cm long but occasionally up to 1.5 cm. Calyx green, outside mostly sparsely pilose, occasionally pilose to villous or arachnoid, inside glandular-hairy; margins often glandular-hairy, lobes basally connate 1–2 mm, subequal, each of the lobes ovate with acute to obtuse apex, dorsal lobe keeled, 4–12 mm long, 3–10 mm wide, entire to erose, dorsal lobe often directed backward below spur, ventral lobes directed forward, lateral lobes directed upward. Corolla 2–4.5 cm long, urceolate with a small limb and a ventral pouch, tube (defined as distance from base of corolla to throat) 2.5 cm long, but distance to tip of pouch 3 cm, widened from 3–6 mm at base to maximally 1.1–1.6 cm (with pouch) in diameter at widest point, bright orange, outside glabrous to sparsely or densely pilose, inside glabrous, markings not seen; spur from nearly absent up to 3 mm long, blunt, straight; throat 5–9 (–13) mm in diameter; limb orange, lobes 1–4 mm long, 2–4 mm wide, subequal, acute-rotund, glabrous. Filaments 13–20 mm long, adnate to base of corolla tube for 3–5 mm, glabrous; anthers ca. 2.5 mm long, 1.8 mm wide. Nectary dorsal, bilobed, 1–2 mm high, densely glandular-tomentose. Ovary 4–6 mm long, 2–4 mm in diameter, glabrous; style 6–8 mm long, glabrous; stigma bilobed. Capsule ca. 7 mm long, 10 mm wide, slightly flattened, bivalved; seeds not seen. Figs. 4A, 12E.

Phenology. Collected in flower from January to December; in fruit in June, August, and November.

Distribution (Fig. 29). Ecuador (Imbabura, Pichincha); near waterfalls and streams or on steep slopes in wet montane forests; (900–) 1500–2100 (–2500) m.

ADDITIONAL SPECIMENS EXAMINED. **Ecuador**. PICHINCHA: Saloya, *Acosta Solís 5782* (F); Biological Station "Río Guajalito," *Blasca 1640* (AAU); old Quito–Santo Domingo rd, Chiriboga, Forest Reserve "La Favorita," *Cerón et al. 7913* (QCNE), *8901* (QCNE); trek from Lloa to Mindo, S side of Río Cinto, *Clark 4494* (QCNE, US); old Quito–Santo Domingo rd, 9 km NE of La Palma, *Croat 38775* (MO); Quito–Santo Domingo rd, *Dodson & Thien 1217* (BH-2 sheets, SEL, US, WIS), *Dodson & Tan 5383* (SEL-2 sheets); 11 km W of Tandapi, trail along Río Chictoa, tributary of Río Pilatón, *Gentry et al. 12125* (AAU, MO, US-2 sheets); old Quito–Santo Domingo rd, Km 59, Floristic-Ecological Reserve "Río Guajalito," *Grijalva 526* (QCA); Quito–Nono–Pacto, between Nono and Tandayapa, *Holm-Nielsen 16132* (AAU, QCA); new Calacalí–Nanegalito rd, Km 20, *Hurtado et al. 1416* (US); Floristic-Ecological Reserve "Río Guajalito," *Jaramillo 7875* (QCA); silver mine above Toachi, *Luer et al. 7381* (SEL); old Quito–Santo Domingo rd shortly after turn from the new Quito–Santo Domingo rd, *Mendoza-T. et al. 610* (AAU, K, MO, QCA, QCNE, SEL, US); Mindo, *Neill & Asanza 10344* (QCNE); Maquipucuna Reserve, ridge N of Cerro de Sosa, ca. 7 km SE of Nanegal, *Webster et al. 27134* (QCA); 6 km NW of Calicalí, Cerro Yanaurcua, *Webster 27274* (US); 5–6 km E of Nanegal, Cerro Campana, *Webster 30038* (AAU, US); "Reserva Río Guajalito," near Chiriboga, *van der Werff et al. 12206* (MO).

Gasteranthus lateralis has been confused mostly with *G. calcaratus* subsp. *oncogastrus*, which also occurs in western Ecuador but generally at lower elevations than *G. lateralis*. The following characteristics in combination set *G. lateralis* apart from *G. calcaratus*: 1) leaf blades obovate to oblanceolate with an obtuse or broadly acute apex, large (usually at least some more than 14 cm long and 8 cm wide), and borne on short petioles (less than 1 cm and mostly approx. 5 mm long); 2) long peduncles (usually exceeding 10 cm); and 3) fairly large urceolate corollas (mostly longer than 3 cm but occasionally only 2 cm long) (Fig. 4A). Virtually each of the characteristics of *G. lateralis* is shared with at least some specimens of the variable *G. calcaratus* (Fig. 6); however, *G. lateralis* is easily distinguished from *G. calcaratus* by its larger stature and the more conspicuous indumentum on the stems and inflorescences. *Gasteranthus macrocalyx* (Fig. 4B) is vegetatively very similar to *G. lateralis*, but differs in various floral characters, e.g., longer, lanceolate calyx lobes (see *G. macrocalyx*, no. 21).

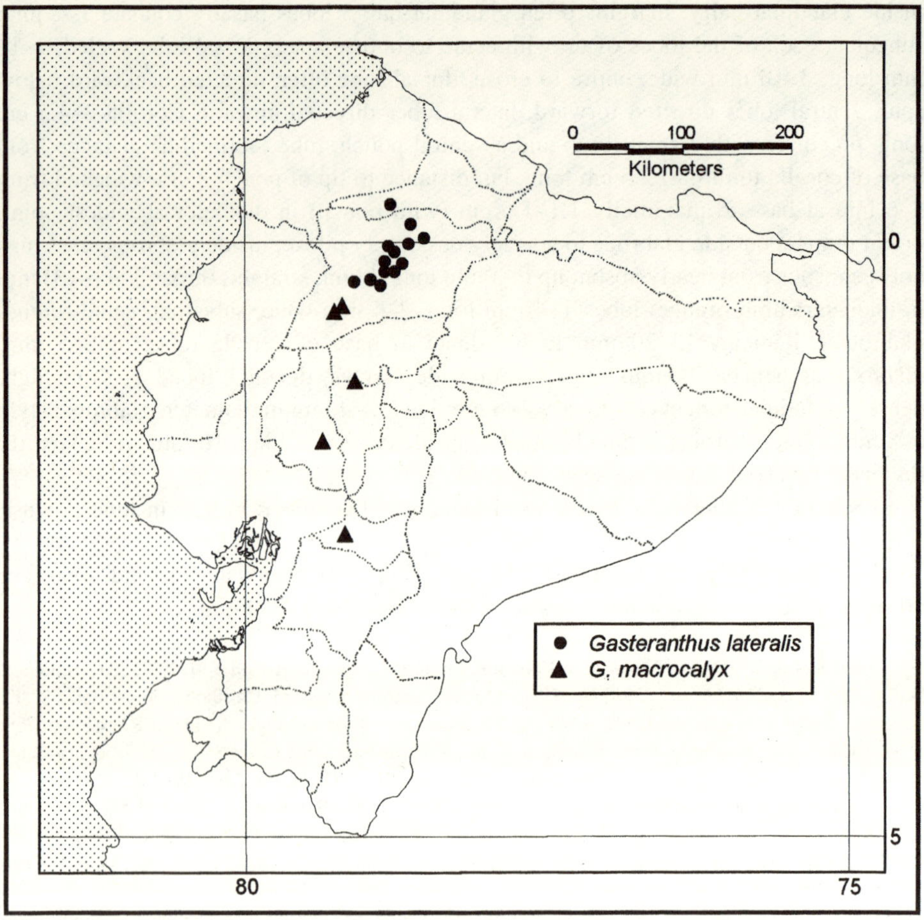

FIG. 29. Distribution of *Gasteranthus lateralis* and *G. macrocalyx*.

20. Gasteranthus leopardus M. Freiberg, Phyton (Horn) 36: 307, fig. 2A. 1996.—TYPE: ECUADOR. Imbabura: Los Cedros Biological Reserve, 1500–1600 m, 29 Jun 1996, *Freiberg 96214* (holotype: QCA!, photo: US!; isotypes: QCNE, ULM!).

Herbs to subshrubs; stems erect, to 1.2 m tall, to 10 mm in diameter, tomentose-villous usually also on older stems but occasionally nearly glabrous, internodes 2–7 cm long, mostly rather irregularly quadrangular, longitudinally striate (striae often hidden by pubescence). Leaves opposite, subisophyllous; blades (4–) 7–22 (–35) cm long, 2–9 (–12) cm wide, elliptic to narrowly elliptic or oblong, base cuneate to acute, apex acute to acuminate, margin serrate, flat, occasionally somewhat rugose along secondary veins, adaxially green to dark green, usually glabrous, sometimes sparsely pilose, abaxially lighter to yellow-green, mostly villous, sometimes pilose or puberulent especially along the veins, lateral veins 9–13 (–20) per side, stomatal clusters not conspicuous; petioles 0.5–2 (–4) cm long, villous-tomentose. Inflorescences open cymes of 1–5 (–15) flowers, pilose to villous; peduncles 2–14 cm long; length of pedicels not exceeding 7 mm, rarely

10–15 mm. Calyx green to brownish, outside pilose to glabrous or glandular, inside pubescent near the middle and glabrous near the edges of the lobes, lobes basally connate 1–2 mm, subequal, each of the lobes narrowly to broadly lanceolate with acuminate apex, 7–20 mm long, 2–7 mm wide, toothed, dorsal lobe directed straight backward below spur, ventral lobes directed forward and appressed below corolla, lateral lobes directed forward and appressed above corolla. Corolla (2.8–) 3.5–5.2 cm long, funnelform with relatively large limb, tube not bent downward, (2.0–) 2.7–3.5 cm, widened from 3–6 mm at base to 7–14 mm in diameter at widest point, yellow, outside pilose and sometimes glandular, inside glabrous or rarely glandular, markings dark purple-brown inside; spur 3–7 mm long, 2–7 mm wide, pointed, straight to slightly bent; throat 7–14 mm in diameter; limb yellow with purple-brown dots/blotches, lobes 4–12 mm long, 5–14 mm wide, subequal, rotund-suborbicular, glabrous. Filaments 13–28 mm long, adnate to base of corolla tube for 3–8 mm, glabrous to sparsely pilose; anthers 2.5–3 mm long, 1.5–1.8 mm wide. Nectary dorsal, bilobed, 0.5–2.5 mm high, pubescent, villous distally, glabrescent basally. Ovary 4–7 mm long, 2–5 mm in diameter, puberulent; style 8–15 mm long, sparsely puberulent to puberulent or pilose; stigma stomatomorphic to weakly bilobed. Capsule ca. 5 mm long, ca. 6.5 mm wide, slightly flattened, bivalved, pubescent; seeds ca. 0.7 mm long, 0.3 mm wide, oblong, papillate. Figs. 9B, 30.

Phenology. Collected in flower from January to April, July to December; in fruit during July.

Distribution (Fig. 40). Colombia (Nariño) and Ecuador (Carchi, Imbabura); montane forests; (1400–) 1700–2100 (–2500) m.

ADDITIONAL SPECIMENS EXAMINED. **Colombia**. NARIÑO: Mpio. Ricaurte, La Planada Natural Reserve, near Chucunes, *Benavides 4070* (US), *10924* (US), *11134* (US), *11203* (US), *Gentry 34980* (COL), *Gentry & Keating 59685* (MO), *Gentry et al. 30550* (COL, MO), *Luteyn & Sylva 13936* (NY, US), *Restrepo CR773* (US), *Restrepo & Ramírez CR575* (US), *Smith & Galeano 1510* (COL), *Stiles 582* (COL); between San Isidro and La Planada, Ricaurte, *Benavides 9234* (US); trail to El Hondón, 6–12 km SW of La Planada, *Gentry et al. 60422* (US); trail from La Planada to Pielapi, *Gentry et al. 63601* (US); Mpio. Mallama, trail Piedrancha–Cabeceras quebrada La Chorrera, *Ramírez P. & Jojoa B. 5751* (US). **Ecuador**. CARCHI: Espejo, Mirador de Golondrinas protected forest, *Clark et al. 2410* (MO, QCNE, SRP, US), *Clark & Dunn 2444* (MO, QCNE, SRP, US), *Clark 2463* (AAU, COL, E, F, G, K, MO, NY, QCA, QCNE, SRP, US, VEN); rd Tulcán–Maldonado, 10 km SE of Maldonado, Campamento Machines, *Harling & Andersson 12334* (SEL, US); Cerro Golondrinas, above division of principal stream, *Hoover 2240* (MO, QCA); above Río Verde, Rafael Quindí's finca, *Hoover & Wormley 1775* (MO); Mira Cantón, El Carmen, rd towards Chical, *Palacios et al. 9852* (QCNE); El Goaltal, slopes of Cerro Golondrinas Hembra, *Palacios & Clark 12427* (QCNE). Maldonado–Tufiño rd, Km 5, *Schwerdtfeger 46* (US).

Gasteranthus leopardus is distinguished by having medium-sized to large (2.8–5.2 cm long) funnelform corollas, which are pale yellow outside but dark purple-brown in the throat, and lanceolate calyx lobes with acuminate apices and a toothed margin (Fig. 9B). The only other species with conspicuously toothed calyx lobes is *G. wendlandianus* (Fig. 9A), and *G. leopardus* may be a close relative of that widely distributed species. *Gasteranthus leopardus* is mainly set apart by having much larger corollas (usually more than 4 cm rather than less than 2.5 cm long, and the throat more than 10 mm rather than less than 7 mm wide).

Gasteranthus leopardus and *G. wendlandianus* are apparently not sympatric (Fig. 40); the latter has never been collected on the western Andean slopes further south than the Colombian department of Cauca (only one collection). It is, however, remarkable that the Colombian collections of *G. leopardus* (which all come from La Planada in Nariño)

FIG. 30. *Gasteranthus leopardus*. A. Habit. B. Inset: section of abaxial leaf surface. C. Corolla. D. Corolla opened to show stamens. E. Calyx. F. Calyx, pistil, and disk. G. Fruit. H. Seeds. (Based on: A, B, *Harling & Andersson 12334*; C–F, *Clark & Dunn 2463*; G, H, *Hoover 2240*.)

are much more variable than the Ecuadorian collections. The latter all have the corollas at least 4 cm long, but in Colombia the length of the corollas varies from 2.8 to 5 cm long, and the smallest-flowered specimens from La Planada thus have only insignificantly larger corollas than the largest corollas recorded from *G. wendlandianus*. A similar dramatic variation in corolla sizes has previously been found in New World Gesneriaceae among varieties of *Kohleria inaequalis* (Benth.) Wiehler, apparently reflecting widespread hybridization between previously isolated species (Kvist & Skog 1992).

21. Gasteranthus macrocalyx Wiehler, Selbyana 2: 78, pl. 24B. 1977.—TYPE: ECUADOR. Los Ríos: Montañas de Ila, Patricia Pilar, 12 km E of Río Palenque Science Center, 650 m, 9 Apr 1977, *Madison 3806* (holotype: SEL!).

Herbs to subshrubs; stems erect, to 1 m tall, to 7 mm in diameter, arachnoid or tomentose distally, internodes 1–3 cm long, quadrangular. Leaves opposite, isophyllous; blades (5–) 10–18 (–22) cm long, 4–8 (–10) cm wide, obovate to oblanceolate, rarely elliptic, base more or less attenuate to acute, apex acute, margin serrate to crenate, flat, adaxially green, glabrous, abaxially lighter green, arachnoid to tomentose at the veins, lateral veins 9–10 per side, stomatal clusters not conspicuous; petioles 0.5–1.5 cm long, arachnoid-tomentose. Inflorescences open to relatively congested cymes of 3–6 flowers, peduncles 7–13 cm long, arachnoid-woolly basally to glabrous and blackish distally; pedicels 0.5–2 cm long, villous. Calyx pale green, arachnoid to woolly outside, glandular-hairy inside, lobes basally connate 3–6 mm, subequal, each of the lobes lanceolate with acuminate apex, 12–22 mm long, 3–7 mm wide, entire, dorsal lobe below gibbous base of the corolla and from there usually directed upward, ventral lobes directed forward-upward usually below corolla, lateral lobes directed upward. Corolla 3–3.5 cm long, urceolate with a relatively small limb and a ventral pouch, tube (defined as distance from base of corolla to throat) 2.5–3 cm, but distance to tip of pouch 3–3.5 cm, widened from 4–7 mm at base to 1.4–2.0 cm (with pouch) in diameter at widest point, yellow-red or orange-red, outside puberulent particularly dorsally, inside puberulent, particularly ventrally, markings not seen; spur virtually absent but base of corolla gibbous distally; throat 6–11 mm in diameter; limb red, lobes 2–4 mm long, 3–5 mm wide, subequal, rotund, sparsely puberulent. Filaments ca. 13 mm long, adnate to base of corolla tube for ca. 5 mm, glabrous; anthers ca. 2 mm long, 1.5 mm wide. Nectary dorsal, semi-annular, up to 1 mm high, tomentose distally, glabrescent basally. Ovary ca. 5 mm long, 4 mm in diameter, puberulent; style ca. 10 mm long, puberulent; stigma stomatomorphic. Capsule and seeds not seen. Fig. 4B.

Phenology. Collected in flower from February to May, and in July and October.

Distribution (Fig. 29). Ecuador (Cañar, Cotopaxi, Los Ríos, Pichincha); premontane to lower montane wet forest; (350–) 600–700 (–1000) m.

ADDITIONAL SPECIMENS EXAMINED. **Ecuador**. CAÑAR: along rd between Azoques and El Triumfo, 1 km S of La Delicia, *Croat 50877* (MO, US).—COTOPAXI: rd Quevedo–El Corazón, 9 km ENE of Moraspungo, *Harling & Andersson 19038* (GB, US).—LOS RÍOS: Montañas de Ila, El Centinela, *Dodson et al. 7535* (MO, SEL); Km 41, Patricia Pilar, *Quelal & Tipaz 350* (QCNE).—PICHINCHA: El Centinela, Patricia Pilar, *Croat 73002* (MO, QCNE); El Centinela, Patricia Pilar, *Dodson 7350* (MO-2 sheets, SEL), *Dodson et al. 14530* (MO, QCNE), *Dodson & Neill 15538* (K, MO-2 sheets, NY, QCNE).

Gasteranthus macrocalyx is characterized by long (12–22 mm), lanceolate calyx lobes, each with an acuminate apex and with a thick and densely appressed indumentum outside (Fig. 4B). The peduncles have a conspicuous indumentum basally, similar to that of the calyces; in contrast, the distal portions of the peduncles and the pedicels are glabrous and darker colored. The inflorescences usually have three or four conspicuous, large (3–3.5 cm long), urceolate corollas (Fig. 4B). The most similar species is *G. lateralis*, also found in western Ecuador but at higher elevations than *G. macrocalyx*. Both species mostly have obovate-oblanceolate leaf blades with a somewhat attenuate and decurrent base and a petiole mostly shorter than 1 cm. The similarity of the two species is also emphasized by the fact that both are glandular-hairy on the inner surface of the calyx lobes, and that both have such short spurs that they may be thought to be dorsally gibbous rather than calcarate. *Gasteranthus lateralis* (Fig. 4A) is set apart from *G. macrocalyx* by its shorter (maximally 12 mm long) ovate calyx lobes with an acute-obtuse apex, and a uniform indumentum on inflorescences and calyces.

The collections of *G. macrocalyx* come from the western foothills of the Andes in the Ecuadorian provinces of Cañar, Cotopaxi, and Los Ríos (near the border with Pichincha). The collection from the province of Cotopaxi (*Harling & Andersson 19083*) is tentatively referred here. It has nearly glabrous inflorescences and only sparsely pilose calyces, but equally large corollas as in typical specimens of *G. macrocalyx*. Nearly all collections are reported to come from 600 to 700 m elevation. A single collection (*Croat 73002*), from La Centinela Ridge on the front range known as Montañas de Ila, is reported to be from 1000 m.

22. Gasteranthus mutabilis L. E. Skog & L. P. Kvist, sp. nov.—TYPE: ECUADOR. Pichincha: old Quito–Santo Domingo road, Km 59, between Chiriboga and Empalme, 1850 m, 3 Nov 1983, *Jaramillo 5965 p.p.* (holotype: QCNE!; isotypes: AAU! K! MO! QCA! US!).

A Gasteranthus orientandino L. E. Skog & L. P. Kvist in lobis calycum ovatis vel ellipticis et in calcaribus leniter angustioribus 2–3 mm × 1–3 mm differt.

Herbs; stems erect, to 60 cm tall, to 5 mm in diameter, pilose to villous with yellow-green trichomes except glabrous just above nodes, internodes 1–3 cm long, terete to quadrangular. Leaves opposite, isophyllous; blades 5–10 cm long, 2–4 cm wide, elliptic, base acute, apex acute, margin serrate, flat, adaxially dark green, pilose, abaxially lighter green, pilose, particularly on the veins, lateral veins 7–9 per side, stomata in conspicuous whitish green clusters; petioles 1–2.5 cm long, pilose. Inflorescences open cymes of 1–2 (–3) flowers, sparsely pilose to glabrous; peduncles 3–8 cm long; pedicels 4–8 mm long. Calyx green, outside pilose, particularly basally, inside minutely puberulent; margins pilose, lobes basally connate ca. 1 mm, subequal, each of the lobes ovate to elliptic, apex acute, dorsal lobe keeled, 8–12 mm long, 3–5 mm wide, subentire, dorsal lobe often directed backward, ventral lobes directed forward and appressed to corolla, lateral lobes directed upward. Corolla 2.2–3.4 cm long, funnelform with large limb, tube 1.0–1.4 cm long, not bent downward, widened from 3–5 mm at base to 8–13 mm in diameter at widest point, red (according to field observations but bright orange-yellow on herbarium sheets), outside sparsely pilose, inside puberulent to velutinous, marked apparently with lines; spur 2–3 mm long, 1–3 mm wide, blunt, straight; throat 8–13 mm in diameter; limb red (yellow on herbarium sheets), lobes unequal, rotund, dorsal lobes 3–7 mm long, 5–12 mm wide, lateral and ventral lobes connate 5–10 mm, lateral lobes 7–9 mm long, 7–15 mm wide, ventral lobe 6–9 mm long, 8–12 mm wide, puberulent to glabrous near margin. Filaments ca. 10–14 mm long, adnate to base of corolla tube for ca. 2 mm, puberulent to

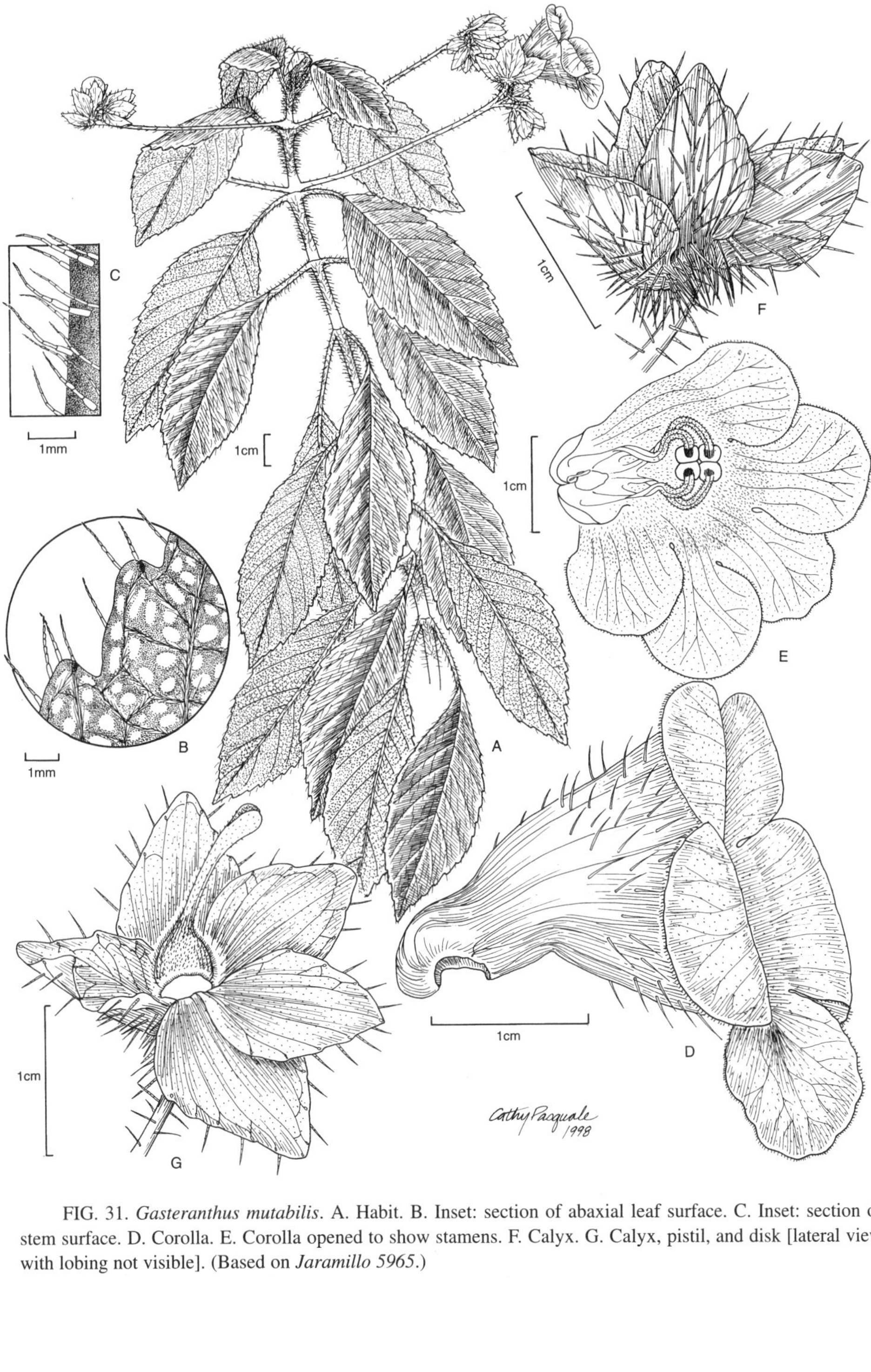

FIG. 31. *Gasteranthus mutabilis*. A. Habit. B. Inset: section of abaxial leaf surface. C. Inset: section of stem surface. D. Corolla. E. Corolla opened to show stamens. F. Calyx. G. Calyx, pistil, and disk [lateral view with lobing not visible]. (Based on *Jaramillo 5965*.)

glabrous; anthers ca. 2 mm long, 1.5 mm wide. Nectary dorsal, broadly and shallowly bilobed, to 1 mm high, sparsely villous. Ovary ca. 4 mm long, 2 mm in diameter, puberulent; style 6–7 mm long, puberulent; stigma stomatomorphic. Capsule and seeds not seen. Figs. 2C, 31.

Phenology. Collected in flower from September to December.

Distribution (Fig. 15). Ecuador (Pichincha); primary and secondary wet montane forests; 1800–2200 m.

Additional Specimens Examined. **Ecuador**. Pichincha: Saloya valley, *Acosta Solís 5770* (F); Palmeras, *Córdova 1* (QCA); Km 59 on old Quito–Santo Domingo rd, Floristic-Ecological Reserve "Rio Guajalito," *Freire F. 463* (GB), *Jaramillo & Grijalva 14528* (QCA), *Rodríguez-Struve 5* (QCA).

The following characteristics identify *G. mutabilis*: 1) a conspicuous villous indumentum on stems and inflorescences; 2) relatively small (up to 10 cm long) elliptic leaves with serrate margins and conspicuous stomatal clusters on the lower surfaces; 3) long-pedunculate, 1- to few-flowered inflorescences, ovate calyx lobes and funnelform corollas with a large limb (Figs. 2C, 31D); and 4) a straight and blunt spur, ca. 3 mm long. A species similar to *G. mutabilis* is the equally rare *G. orientandinus* from the eastern Andean slopes of Ecuador, but the latter species has flowers that differ by having narrowly lanceolate calyx lobes (Fig. 2D) and a longer and wider spur (6–8 mm long). The vegetatively most similar species, however, is *G. extinctus*, but this species has very different urceolate corollas (Fig. 5B).

As dried specimens, the corollas of *G. mutabilis* appear bright orange-yellow. According to the labels on several vouchers, some plants have reddish corollas ("fls. red," "flores tomates," "fuchsia oscuro," "corolla roja," etc.), whereas others of different collections give the color as orange. This may indicate that the corolla color changes during the process of drying, or that populations have plants with differently colored corollas on different plants. This latter phenomenon can be seen in some populations of *G. crispus*, which have some plants with yellow-orange corollas and other plants with red corollas. Apparently, seeds from the same capsule can result in plants with either corolla color.

Note: Some specimens of *G. glaber* have been distributed (in QCA and US) with the same number as the type of *G. mutabilis* (*Jaramillo 5965*), and the two species are mixed on the isotype sheet at QCA.

23. Gasteranthus orientandinus L. E. Skog & L. P. Kvist, sp. nov.—Type: Ecuador. Morona-Santiago: Río Paute, 2–3 km W of Mendez, 1900 ft, 13 Dec 1944, *Camp E-1492* (holotype: QCA!; isotypes: NY! S! SEL! US!).

A Gasterantho mutabili L. E. Skog & L. P. Kvist lobis calycum anguste lanceolatis 8–14 mm × 2–4 mm et a Gasterantho herbaceo L. E. Skog & L. P. Kvist in inflorescentiis peduncularibus differt.

Herbs; stems apparently erect, ca. 50 cm tall, to 3 mm in diameter (from limited material), glabrescent to tomentose distally, internodes 1–2 cm long, terete. Leaves opposite, subisophyllous; blades 6–11 cm long, 2.5–4.5 cm wide, oblique, ovate to elliptic, base cuneate to acute, apex acute to acuminate, margin weakly serrate, flat, adaxially dark green (and in some plants finely mottled with white), glabrous, abaxially lighter green, glabrous apart from puberulent-tomentose primary and secondary veins, lateral veins 5–6 per side, stomata in conspicuous but lighter clusters; petioles 1–4 cm

FIG. 32. *Gasteranthus orientandinus*. A. Habit. B. Inset: section of abaxial leaf surface [stomata not shown]. C. Corolla. D. Corolla opened to show stamens. E. Calyx. F. Calyx, pistil, and disk. G. Fruit. H. Seeds. (Based on *Camp E-1492*.)

long, tomentose-villous. Inflorescences open cymes of 4–6 flowers, pilose; peduncles 3–4 cm long; pedicels 1–2 cm long. Calyx green, pilose to villous, lobes basally connate ca. 2 mm, subequal, each of the lobes narrowly lanceolate with acuminate apex, centrally with a darker longitudinal keel, 8–14 mm long, 2–4 mm wide, subentire, dorsal lobe directed backward appressed below spur, ventral lobes directed forward, lateral lobes directed upward. Corolla 3.5–4 cm long, funnelform with large limb, tube 1.5–2.0 cm long, not bent downward, widened from 3–4 mm at base to 10–15 mm in diameter at widest point, yellow, outside pilose to villous mainly on the spur, inside glandular-puberulent particularly in the throat, marked apparently with lines; spur 6–8 mm long and 4–6 mm wide, straight or directed somewhat downward, apex blunt; throat 10–15 mm in diameter; limb yellow, lobes 5–9 mm long, 4–9 mm wide, subequal, rotund, glabrous. Filaments 9–14 mm long, adnate to base of corolla tube for 2–3 mm, glabrous to glandular-puberulent; anthers ca. 2 mm long, 1.5 mm wide. Nectary dorsal, bilobed, ca. 1 mm high, appressed glandular-pilose. Ovary ca. 3 mm long, 2 mm in diameter, densely puberulent; style 7 mm long, glabrous; stigma stomatomorphic. Capsule ca. 6 mm long, ca. 8 mm wide, flattened, bivalved, splitting somewhat irregularly; seeds ca. 0.3 mm long, 0.2 mm wide, globose, papillate. Figs. 2D, 32.

Phenology. Collected in flower and fruit during December.

Distribution (Fig. 15). Ecuador (Morona-Santiago); on shady cliff-sides of small streams in wet forests; 600 m.

Gasteranthus orientandinus is set apart from all other species by having relatively large (3.5–4 cm long) funnelform corollas with a wide throat and limb, and a straight and blunt spur, 6–8 mm long (Fig. 2D). The most similar species may be *G. mutabilis* and *G. herbaceus* from northwestern Ecuador and Colombia, respectively. The former (Fig. 2C) has a much shorter spur (maximally 3 mm long), and the latter (Fig. 3E) a downward-bent and usually considerably larger spur. *Gasteranthus orientandinus* also differs from *G. mutabilis* by having narrowly lanceolate rather than ovate to elliptic calyx lobes, and from *G. herbaceus* by having pedunculate rather than epedunculate inflorescences. *Gasteranthus orientandinus* is vegetatively fairly similar to *G. mutabilis*, suggesting that the two species may be closely related.

Gasteranthus orientandinus is known only from the type collection. The name reflects that it is the only species of *Gasteranthus* that is endemic to the eastern Ecuadorian region known as "Oriente."

24. Gasteranthus osaensis L. E. Skog & L. P. Kvist, sp. nov.—TYPE: COSTA RICA. Puntarenas: Cantón de Osa, vicinity of Jalaca station, between Río Esquinas and Palmar Sur de Osa, 30 m, 10 May 1950, *Allen 5533* (holotype: US!; isotypes: AAU! LAM! NY!).

A Gasterantho acropodo (J. D. Smith) Wiehler corollis minoribus 2–2.5 cm longis ventraliter conspicue sacculatis tubo ± valde urceolato conspicue piloso et inflorescentiis 3–10-floris differt.

Herbs; stems erect, to 1.5 m tall, to 6 mm in diameter, sparsely pilose to tomentose distally, internodes 3–7 cm long, quadrangular. Leaves opposite, subisophyllous; blades 8–20 mm long, 3–8 cm wide, oblique, ovate to elliptic, base oblique and attenuate, apex acute to acuminate, margin weakly serrate to subentire, flat, adaxially green, sparsely pilose to glabrous, abaxially lighter green, puberulent to tomentose, mainly on veins, lateral veins 8–9 per side, stomatal clusters not conspicuous; petioles 2–3 cm long, pilose to

FIG. 33. *Gasteranthus osaensis*. A. Habit. B. Inset: section of abaxial leaf surface. C. Corolla. D. Corolla opened to show stamens. E. Calyx. F. Calyx, pistil, and disk. G. Fruit. H. Seeds. (Based on *Allen 5533*.)

tomentose. Inflorescences open cymes of 3–10 flowers, glabrous to sparsely pilose; peduncles 4–6 cm long; pedicels 0.5–3 cm long. Calyx green, pilose, lobes basally connate 1–3 mm, subequal, each of the lobes ovate to lanceolate with acute apex, 5–10 mm long, 4–7 mm wide, entire, pilose, dorsal lobe directed backward and surrounding spur, ventral lobes directed forward, lateral lobes directed upward. Corolla 2.0–2.5 cm long, ventricose to urceolate with a relatively large limb and a ventral pouch, tube 1.4–2.0 cm long, widened from 3–4 mm at base to maximally 10–13 mm (with pouch) in diameter at widest point, yellow, outside pilose, trichomes sometimes glandular, inside glandular near base, glandular-puberulent in throat, markings not seen; spur 3–5 mm long, diameter up to 3 mm, apex blunt, directed downward; throat 5–8 mm in diameter; limb yellow, lobes 2–3 mm long, 3–4 mm wide, equal, rotund, glabrous. Filaments 12–16 mm long, adnate to base of corolla tube for 4–6 mm, puberulent below, glabrous above; anthers ca. 1.5 mm long, 1.5 mm wide. Nectary dorsal, bilobed, to 1.5 mm high, densely villous to glabrescent. Ovary 2–3 mm long, 3–4 mm in diameter, puberulent; style 6–10 mm long, glabrous; stigma stomatomorphic to weakly bilobed. Capsule ca. 3 mm long, 6 mm wide, flattened, bivalved; seeds ca. 0.2 mm long, 0.1 mm wide, oblong, striate. Figs. 10A, 33.

Phenology. Collected in flower from March to August and in December; in fruit only in June.

Distribution (Fig. 20). Costa Rica (Puntarenas: Osa peninsula); on heavily shaded streamsides and steep slopes in wet forest; 0–250 m.

ADDITIONAL SPECIMENS EXAMINED. **Costa Rica**. PUNTARENAS: Osa Peninsula, upper Aguabuena, 5 km W of Rincón, *Thomsen 374* (C); 4 km W of Rincón de Osa, *Burger & Stolze 5481* (F); 5 km W of Rincón de Osa, *Burger & Gentry 9015* (NY); Osa Peninsula, 6 km W of Rincón, *Grant & Rundell 92-01921* (F, US); Forest reserve Golfo Dulce, Rancho Quemado, rd between Rancho and Drake, *Hammel & Nepokroeff 18256* (US); Rancho Quemado, *Marín 86* (MO), *481* (US); 4 km W of Rincón de Osa, *Raven 21609* (F); between Golfo Dulce and Río Térraba, *Skutch 5381* (US); Rincón de Osa, W of airstrip, *Utley & Utley 1112* (F).

Gasteranthus osaensis is the only species of *Gasteranthus* found on the Osa peninsula of Costa Rica. It is similar to *G. acropodus*, which is mainly distributed along the Caribbean coast of Panama but also occurs in the province of Limón in eastern Costa Rica. *Gasteranthus osaensis* (Fig. 10A) differs from *G. acropodus* (Fig. 9C) by having smaller corollas (2–2.5 cm rather than 2.5–4.5 cm long) with a conspicuous ventral pouch and more or less strongly urceolate rather than ventricose tubes, which are conspicuously pilose rather than more or less appressed-pilose outside, and by having inflorescences with more flowers (mostly more than 5 rather than fewer than 5). *Gasteranthus osaensis* may also have affinities with *G. calcaratus*, which is common in Colombia and Ecuador. The corollas of the latter species (Fig. 6) usually differ by having an inflated pouch protruding beyond the throat, and by being glabrous or having an appressed indumentum outside. A third somewhat similar species, *G. timidus* (Fig. 8D) from western Ecuador, differs from *G. osaensis* by having more congested inflorescences (pedicels less than 5 mm long) and smaller corollas (1.1–1.7 mm long) with a densely villous indumentum outside.

25. Gasteranthus otongensis M. Freiberg, Phyton (Horn) 38: 170, fig 2. 1998.—TYPE: ECUADOR. Cotopaxi: Bosque Protector Otonga, 27 Jul 1996, *Freiberg 96263* (holotype: QCA; isotypes: QCNE, ULM).

Subshrubs; stems erect, to 1 m tall, to 4–8 mm in diameter, densely villous-tomentose toward apex, internodes 2–5 cm long, terete, usually unbranched. Leaves opposite, isophyllous; blades 9–12 cm long, 4–6 cm wide, elliptic, base cuneate, apex obtuse to shortly acute, margin crenate, adaxially dark green, villous to tomentose, abaxially sea-green (?), villous to tomentose on the veins, lateral veins 7–11 per side, stomatal clusters not conspicuous (?); petioles 1–2.5 cm long, densely villous-tomentose. Inflorescences open subumbellate cymes of 6–20 flowers, densely villous-tomentose; peduncles 3–8 cm long; pedicels 3–6 (–8) cm long. Calyx red, densely villous-tomentose on both sides, lobes basally connate for 4 mm, subequal, each of the lobes lanceolate, apex acuminate, dorsal lobe 23 mm long, 6 mm wide, ventral lobes 27–30 mm long, 6–7 mm wide, lateral lobes 27 mm long, 7 mm wide, denticulate to subentire, dorsal lobe curved upward surrounding spur. Corolla nearly hidden by the calyx, ventricose with a relatively large limb, the ventral pouch ca. 2 mm long, tube 1.8–2.2 cm long, widened from 4 mm at base to ca. 8 mm in diameter at widest point, orange-yellow to orange-red from yellow base, outside villous-tomentose with reddish hairs toward apex, inside papillate near base, slightly tomentose near throat, markings not seen; spur 2 mm long, blunt, straight; throat not contracted; limb terminal, lobes 2–3 mm long, 3 mm wide, subequal, rotund, entire. Filaments ca. 10–11 mm long, adnate to base of corolla tube for ca. 4 mm, glabrous above a papillate base; anthers 2.5 mm long, 2 mm wide; staminode 3 mm long. Nectary annular (?), dorsally bilobed, 1.5 mm high, sparsely tomentose. Ovary 4 mm long, 2 mm in diameter, ovoid, laterally compressed, tomentose; style 6 mm long, tomentose; stigma bilobed. Capsule bivalved, flattened; seeds black.

Phenology. Collected in flower and fruit during July.

Distribution (Fig. 23). Ecuador (Cotopaxi); in shaded areas near small streams in the valley of the Río Esmeraldas in primary montane forest; 1800 m.

Gasteranthus otongensis is most similar to *G. glaber*, which is equally distributed in upper montane cloud forest on the western Andean slopes, but further to the north, from the Ecuadorian province of Pichincha north to the Colombian department of Nariño. Both species have red calyx lobes that nearly hide the relatively small corollas, and both often have leaves with a particularly conspicuous venation. *Gasteranthus glaber* (Fig. 7C) differs from *G. otongensis* in being nearly glabrous, in having semiorbicular to ovate rather than lanceolate calyx lobes, and in having corollas with a contrasting black rather than a red limb.

Type specimens were requested from ULM, QCA, and QCNE, but have not yet been received or examined. The description was derived from the original publication and illustration.

26. Gasteranthus pansamalanus (J. D. Smith) Wiehler, Selbyana 1: 155. 1975. *Besleria pansamalana* J. D. Smith, Bot. Gaz. 16: 197, pl. 17. 1891.—TYPE: GUATEMALA. Alta Verapaz: Pansamalá, 3800 ft, Jan 1887, *Türckheim 196* (lectotype, designated by Morton, 1939: US!; isolectotypes: GH, NY!).

Episcia aurea T. S. Brandegee, Univ. Calif. Publ. Bot. 6: 374. 1917.—TYPE: MEXICO. Chiapas: Cerro del Boquerón, Sep 1913, *Purpus 7533* (lectotype, here designated: UC!).

Besleria pansamalana var. *ecuadorensis* Fritsch, Notizbl. Bot. Gart. Berlin-Dahlem 11: 975. 1934. *Besleria ecuadorensis* (Fritsch) C. V. Morton, Contr. U.S. Natl. Herb. 26: 471. 1939. *Gasteranthus ecuadorensis* (Fritsch) Wiehler, Selbyana 1: 154. 1975.—TYPE: ECUADOR. Pichincha: Niebly, *Sodiro 119/50* (holotype: B, destroyed, photo: F!; lectotype, here designated: P!; isolectotype: QPLS, photo: US!).

Gasteranthus magentatus M. Freiberg, Phyton (Horn) 38: 168, fig 1. 1998.— TYPE: ECUADOR. Cotopaxi: Bosque Protector Otonga, 1800 m, 5 Apr 1996, *Freiberg 96047* (holotype: ULM).

Herbs to subshrubs; stems erect, to 1.5 m tall (but one collection said to be 4 m tall), to 6 mm in diameter, glabrescent to arachnoid distally, internodes 1–9 cm long, mostly quadrangular, sometimes terete, cortex somewhat longitudinally wrinkled, unbranched. Leaves opposite, isophyllous; blades 5–18 cm long, 3–9 cm wide, ovate to elliptic or obovate, base acute, apex acute, occasionally obtuse, margin serrate to subentire, flat, adaxially green, appressed pilose to glabrous, abaxially lighter green, arachnoid to villous or pilose along veins, lateral veins 5–7 (–10) per side, stomatal clusters often not conspicuous; petioles 0.5–2 cm long, arachnoid-tomentose. Inflorescences reduced pedunculate cymes of 1 flower, arachnoid (rarely glabrous); peduncles 2–8 (–10) cm long; pedicels 1–5 (–12) mm long. Calyx green, outside glabrous to arachnoid or appressed-pilose, usually with white blotches distally, inside glabrous, lobes basally connate 1–2 (–5) mm, equal to subequal, each of the lobes ovate to lanceolate with acute to acuminate or obtuse apex, dorsal lobe keeled, 5–14 (–23) mm long, 3–10 (–15) mm wide, subentire, dorsal lobe directed backward and upward, ventral lobes directed forward, lateral lobes directed upward. Corolla 2.2–3.6 cm long, urceolate with a relatively small limb and an inflated ventral pouch, tube (defined as distance from base of corolla to throat) 1.1–1.9 cm long, but distance from base of the corolla to tip of pouch 1.8–3.2 cm, widened from 3–6 mm at base to maximally 0.9–1.7 cm (with pouch) in diameter at widest point, yellow-orange or reddish orange, outside appressed pilose to glabrous, inside glabrous, markings not seen; spur 2–5 mm long, blunt, straight; throat 3–7 (–10) mm in diameter; limb yellow-orange, lobes 2–3 mm long, 3–4 mm wide, subequal to unequal, rotund-ovate, glabrous but glandular. Filaments 9–17 (–20) mm long, adnate to base of corolla tube for 2–6 mm, glabrous; anthers ca. 2 mm long, 1.5–3 mm wide; staminode 5 mm long. Nectary thickened dorsally, bilobed to semi-annular, 1–1.5 mm high, appressed pubescent. Ovary 5–7 mm long, 4–5 mm wide, sparsely puberulent-glabrous; style 5–12 mm long, sparsely puberulent to glabrous; stigma stomatomorphic to weakly bilobed. Capsule ca. 10 mm long, 10–15 mm wide, somewhat flattened, bivalved; seeds 0.5 mm long, 0.2 mm wide, ellipsoid, striate. Figs. 5D, 12A.

Phenology. Collected in flower from January to April and June to December; in fruit in January, March, and April.

Distribution (Fig. 34). Mexico (Chiapas), Guatemala (Alta Verapaz, Baja Verapaz, Chimaltenango, Quezaltenango, Quiché, Retalhuleu, San Marcos, Suchitepéquez), Colombia (Cauca, Huila, Nariño, Putumayo), and Ecuador (Azuay, Bolívar, Carchi, Cotopaxi, Imbabura, Morona-Santiago, Napo, Pastaza, Pichincha, Sucumbíos, Tungurahua, Zamora-Chinchipe); on shaded steep banks or streamsides in dense, wet montane forests; (400–) 1700–2400 (–3800) m.

REPRESENTATIVE SPECIMENS. **Mexico**. CHIAPAS: above Talquian, Volcán Tacaná, *Breedlove & Smith 31633* (DS), *Breedlove 42568* (CAS); hills E of Unión Juárez, lower slopes of Volcán Tacaná, *Miller et al. 2654* (MO, US); Cerro del Boquerón, *Purpus 6850* (UC).

Guatemala. ALTA VERAPAZ: rd between Tactic and the divide on rd to Tamahú, *Standley 90664* (F, US).—BAJA VERAPAZ: Chilasco, on Concepción rd, 6 km E, *Contreras 10891* (US).—CHIMALTENANGO: Calderas, *Johnston 1804* (F).—QUEZALTENANGO: Chiquihuite, *Standley 68145* (F-2 sheets, US).—QUICHÉ: Cerro Putul,

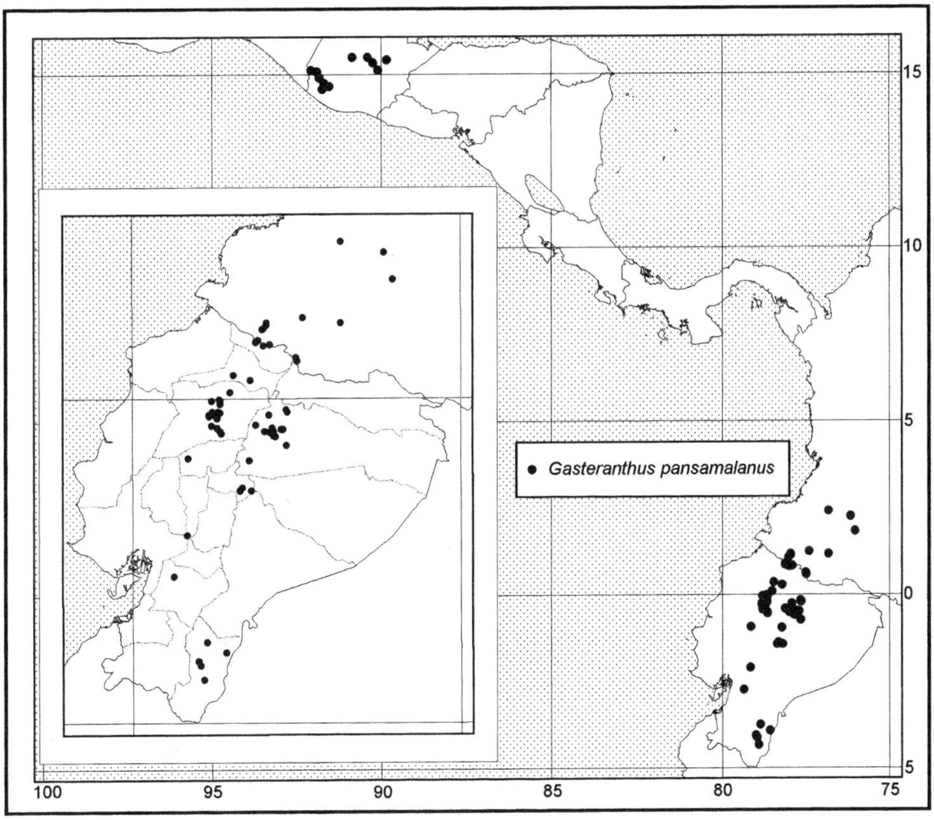

FIG. 34. Distribution of *Gasteranthus pansamalanus*. Inset: Distribution of *G. pansamalanus* in Ecuador and southwestern Colombia.

Skutch 1827 (GH, US).—RETALHULEU: Finca Helvetia, *Muenscher 12440* (F).—SAN MARCOS: near Aldea Fraternidad, between San Rafael Pie de la Cuesta and Palo Gordo, *Williams et al. 26235* (F, NY, US), *26270* (F).—SUCHITEPÉQUEZ: vic. Finca Montecristo, SE of Santa María de Jesús, *Steyermark 35201* (F).

Colombia. CAUCA: W of Tambo, *Haught 5195* (US).—HUILA: Valley of Río Cedro, SE of Pitalito, *Schultes & Villareal 5220* (COL, GH, US).—NARIÑO: La Planada, 7 km above Ricaurte, *Gentry et al. 34988* (COL, MO, US).—PUTUMAYO: Río Minchoy valley, trail from San Francisco de Sibundoy to Minchoy, *Londoño & Lozano 224* (AAU, TULV, US). **Ecuador**. AZUAY: Cuenca-Cola de San Pablo rd, Contego Encampment, *Boeke & Loyola 1020* (NY, SEL).—BOLÍVAR: Hda. Changuil, *Cornejo & Bonifaz 4245* (US).—CARCHI: Maldonado, Quebrada Naranjo near waterfall, *Dorr & Barnett 6097* (NY, QCA); S of Tulcán, between Santa Bárbara and La Bonita, *Molau et al. 2625* (AAU, GB, QCA).—COTOPAXI: Cantón Pilalo, Cerro Puchuato, *Cerón & Villavicencio 2806* (MO, US-2 sheets).—IMBABURA: Intag valley above Apuela, *Holm-Nielsen & Jaramillo 23331* (AAU).—MORONA-SANTIAGO: Cordillera de Cutucú, above Río Tzontza, *Prieto CuP-13* (NY).—NAPO: above upper Río Borja, slopes of Guagra Urcu, *Holm-Nielsen et al. 26919* (AAU, QCA), *27103* (AAU, QCA); 1 km SW of Baeza, *Øllgaard & Balslev 10181* (AAU).—PASTAZA: 5 km W of Río Negro, *Dodson & Thien 1906* (US).—PICHINCHA: old Quito–Santo Domingo rd, Km 59, Las Palmeras, *Borchsenius 91408* (AAU, QCA, QCNE); Bellavista Lodge, near Mindo, Nambillo Reserve, *Clark et al. 3822* (QCNE, US); Cordillera de los Yumbos, *Clark 4595* (AAU, COL, MO, QCA, QCNE, US); old Quito–Santo Domingo rd, above Chiriboga, *Mendoza-T. et al. 622* (QCA, QCNE, US).—SUCUMBÍOS: Santa Bárbara–La Bonita rd, *Harling & Andersson 12496* (GB, SEL, US).—TUNGURAHUA: Baños–Puyo rd, confluence of Río Pastaza and Río Negro, *Kvist 60318* (AAU, QCA, US); Cashurco near Río Negro, *Lugo S. 809* (GB, SEL, US).—ZAMORA-CHINCHIPE: Zamora–Loja rd, Km 5, *Harling & Andersson 23933* (GB, QCA, US); Quebrada del León, affluent of Río Bombuscara S of Zamora, *Madsen & Knudsen 86815* (AAU, MO, QCA).

Gasteranthus pansamalanus is easily distinguished by having one-flowered inflorescences with medium-sized (2.2–3.6 cm long) strongly urceolate corollas with a protruding and conspicuously ridged ventral pouch (Fig. 5D). The inflorescences are pedunculate, but they never develop more than one flower. *Gasteranthus perennis* is only other species with urceolate corollas (Fig. 4C) that consistently has one-flowered pedunculate inflorescences; it is easily recognized by its relatively narrow leaves (0.8–3.2 cm wide). The leaves of *G. pansamalanus* are mostly elliptic or nearly so and 3–9 cm wide; they are acute apically and basally, and mostly 10–15 cm long.

Gasteranthus pansamalanus has a remarkably disjunct distribution. It is widespread in the montane forests of Ecuador and southern Colombia, where it occurs on both the eastern and the western slopes of the Andes Mountains. The species also occurs widely in Guatemala and in Mexico in the state of Chiapas near the border with Guatemala. The disjunct South American and Central American populations of *G. pansamalanus* are surprisingly similar. The latter may have on the average somewhat larger leaves and flowers, but there are no consistent recognizable differences. This may indicate that the Central American populations have been established relatively recently perhaps following long-distance dispersal from South America, similar to the pattern noted by Smith (1994) in *Columnea nervosa* (Klotzsch ex Oersted) Hanstein.

Freiberg (1998) described a plant with a reddish corolla as a distinct species, *Gasteranthus magentatus*. In all other respects, his species resembles *G. pansamalanus*. Populations of some species of *Gasteranthus* are not unusual in having plants with different corolla colors, especially yellow-orange or red.

27. Gasteranthus perennis (C. V. Morton) Wiehler, Selbyana 1: 155. 1975. *Besleria perennis* C. V. Morton, Contr. U.S. Natl. Herb. 26: 470. 1939.—TYPE: ECUADOR. Cotopaxi: Cantón Pujilí, Hacienda Salento, 1000 m, 20 Nov 1934, *Mexia 6719* (holotype: US!).

Herbs; stems decumbent or erect, rising to at least 60 cm tall, to 5 mm in diameter, glabrescent to arachnoid distally, internodes 0.3–2.5 cm long, terete. Leaves both opposite and ternate on the same shoots, isophyllous; blades 4–13 cm long, 0.8–3.2 cm wide, narrowly lanceolate to narrowly oblanceolate, base narrowly cuneate, apex acute to acuminate, margin remotely crenate to sinuate, flat, adaxially dark green, glabrous, abaxially lighter green or purplish, particularly on veins, arachnoid to villous, mainly on the veins, lateral veins 5–7 per side, stomatal clusters conspicuous; petioles 0.3–1.5 cm long, arachnoid. Inflorescences pedunculate cymes of 1 flower, glabrous; peduncles 2–6 cm long; pedicels 2–5 mm long. Calyx dark green to purplish, arachnoid, lobes basally connate 1–2 mm, subequal, each of the lobes narrowly lanceolate with a long-acuminate apex, 12–24 mm long, 2–6 mm wide, entire, dorsal lobe directed backward below spur and from there upward, ventral lobes directed upward, lateral lobes directed upward. Corolla 2.5–3 cm long, urceolate with a large limb and a ventral pouch surpassing the throat, tube (defined as distance from base of the corolla to throat) 1.4–1.8 cm long but distance from base to tip of pouch 2.2–2.8 cm, widened from 3–6 mm at base to maximally 1.6–2.0 cm (with pouch) in diameter at widest point, bright red, outside puberulent to arachnoid, inside puberulent to arachnoid mainly towards the throat, markings not seen; spur 2–3 mm long, blunt, straight; throat 6–15 mm in diameter; limb red, lobes 3–4 mm long, 5–8 mm wide, subequal, rotund, glabrous. Filaments 11–17 mm long, adnate to base of corolla tube for 3–5 mm, puberulent; anthers ca. 2.5 mm long, 1.8 mm wide. Nectary dorsal, bilobed,

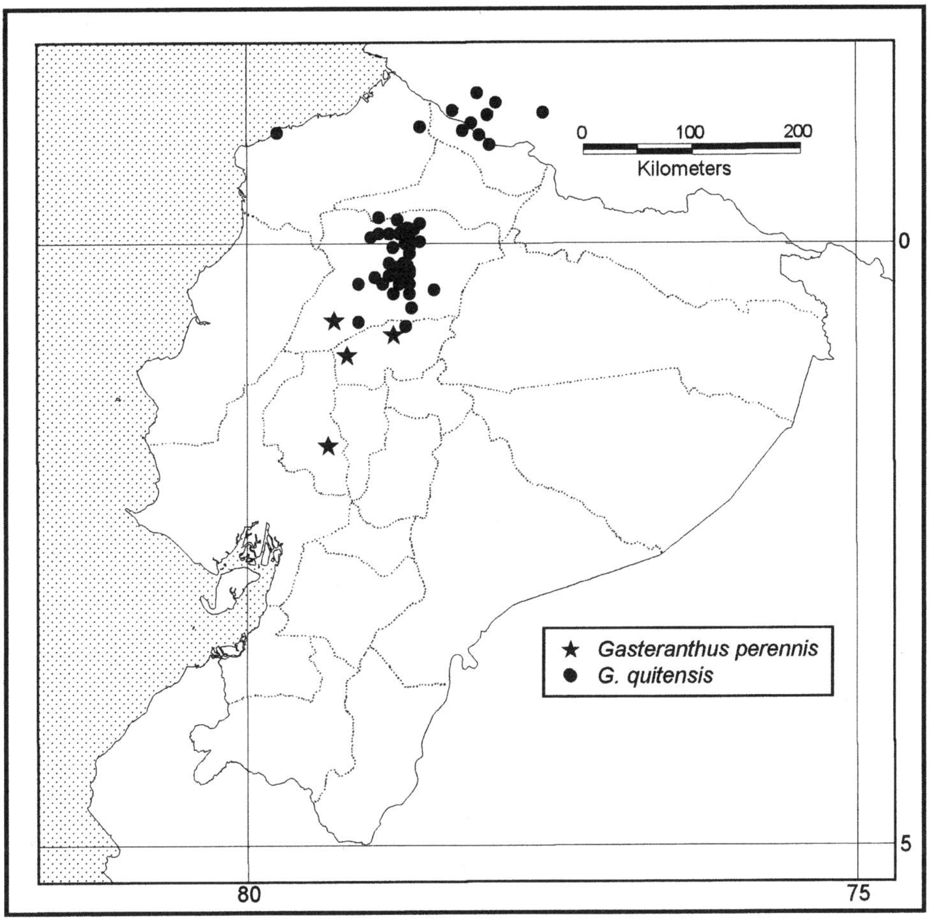

FIG. 35. Distribution of *Gasteranthus perennis* and *G. quitensis*.

ca. 1 mm high, villous-tomentose. Ovary ca. 4 mm long, 2.5 mm in diameter, velutinous; style 7–9 mm long, puberulent; stigma stomatomorphic. Capsule ca. 3 mm long, 6 mm wide, flattened, bivalved and secondarily 4-valved; seeds ca. 0.4 mm long, 0.3 mm wide, globose, striate. Figs. 4C, 11C.

Phenology. Collected in flower from February to April and in November; in fruit only in April.

Distribution (Fig. 35). Ecuador (Cotopaxi, Los Ríos, Pichincha); in deep shade on steep slopes in dense cloud forests; 600–1000 m.

ADDITIONAL SPECIMENS EXAMINED. **Ecuador**. COTOPAXI: Quevedo–Latacunga rd, Km 46, *Holm-Nielsen 2899* (AAU, F, MO, NY, US); foothills above Valencio, *Mathias & Taylor 5190* (UCLA).—LOS RÍOS: Río Pita, Hacienda Clementina, Cerro Mombe, *Asplund 5523* (AAU, F, GB, US).—PICHINCHA: rd Patricia Pilar to Montañas de Ila, Km 18, N side of Torre de Bijagual, *Øllgaard & Borchsenius 100682* (AAU).

Most specimens of *G. perennis* are very distinctive by having narrowly lanceolate leaf blades (4–13 cm long), which are 5–6 times longer than wide. The crenate margin makes the blades appear nearly lobate. In addition, the plants mostly have three rather than two opposite leaves at a node. The lower leaf surfaces show very conspicuous stomatal clusters. Some specimens, however, have leaves that are only ca. 3.5 times longer than wide, but floral characters also help to distinguish *G. perennis* (Fig. 4C). The calyx lobes of *G. perennis* are narrowly lanceolate with a long narrow acuminate apex (only *G. macrocalyx*, Fig. 4B, has similar calyx lobes). The urceolate corollas (Fig. 4C) vary considerably in the diameter of the throat (from 6–15 mm). Although the inflorescences are pedunculate, *G. perennis* apparently never develops more than one flower per inflorescence. Only one other species, *G. pansamalanus*, always has pedunculate, 1-flowered inflorescences, but the other characters cited easily distinguish *G. pansamalanus* from *G. perennis*.

28. Gasteranthus quitensis Bentham, Pl. Hartweg. 233. 1846. *Besleria quitensis* (Bentham) Hanstein, Linnaea 34: 334. 1865.—TYPE: ECUADOR. Pichincha: Nanegal, *Hartweg s.n.* (holotype: K!).

Besleria sodiroana Fritsch, Repert. Spec. Nov. Regni Veg. 18: 12. 1922.—TYPE: ECUADOR. Pichincha: San Florencio, 1600 m, *Sodiro 119/53* (holotype: B, destroyed; lectotype, here designated: QPLS, photo: US!).

Herbs to subshrubs; stems erect, to 1 m tall, to 6 mm in diameter, thickened at nodes, pilose to villous, internodes 1–3 (–7) cm long, mostly terete, occasionally quadrangular, cortex wrinkled. Leaves opposite, subisophyllous; blades 4–14 cm long, 2–6 cm wide, ovate to elliptic, base usually cuneate, occasionally obtuse, apex obtuse to acuminate, margin serrate, often weakly rugose between secondary veins, adaxially dark green, pilose mainly in bands between secondary veins, abaxially lighter green, pilose, mainly on the veins, lateral veins 13–18 per side and distance between adjacent veins usually less than 5 mm and never exceeding 7 mm, intersecondary veins not conspicuous, stomatal clusters not conspicuous; petioles 1–6 cm long, pilose to villous. Inflorescences relatively congested cymes of 3–12 flowers, glabrous, occasionally sparsely pilose; peduncles 3–8 cm long; pedicels 1–8 mm long. Calyx light green, outside pilose, inside glabrous, lobes basally connate 1–2 mm, subequal, each of the lobes ovate to lanceolate, apex acute to acuminate, dorsal lobe keeled, 6–10 mm long, 2–5 mm wide, serrate, erose or subentire, usually pilose, dorsal lobe often directed backward surrounding spur, ventral lobes usually directed forward, lateral lobes directed upward. Corolla 1.2–2.5 cm long, urceolate with a relatively small limb and a ventral pouch, tube length (defined as distance from base of the corolla to throat) 0.7–1.4 cm, but length from base to tip of pouch 1.1–1.9 cm long, widened from 2–3 mm at base to maximally 6–17 mm (with pouch) in diameter at widest point, yellow-orange-red, outside puberulent, inside puberulent in pouch, markings none; spur 1–3 mm long, blunt, straight; throat 3–6 mm in diameter; limb orange-red, lobes 1.5–3.5 mm long, 1.5–4.5 mm wide, subequal to unequal, rotund to acute, glabrous. Filaments ca. 7–9 mm long, adnate to base of corolla tube for ca. 2 mm, glabrous; anthers ca. 1.5 mm long, 1.5 mm wide. Nectary dorsal, bilobed, ca. 0.5 mm high, glabrous. Ovary 2–3 mm long, 2–4 mm in diameter, glabrous to puberulent; style 4–6 mm long, glabrous; stigma stomatomorphic to weakly bilobed. Capsule ca. 3 mm long, ca. 6 mm wide, flattened, bivalved; seeds ca. 0.3 mm long, 1.5 mm wide, elliptic. Figs. 4D, 11F.

Phenology. Collected in flower in all months of the year; in fruit during February and December.

Distribution (Fig. 35). Colombia (Nariño) and Ecuador (Carchi, Cotopaxi, Esmeraldas, Imbabura, Pichincha); along shaded streams, on slopes, and near waterfalls in wet forests; (0–) 600–2400 (–2700) m.

REPRESENTATIVE SPECIMENS (from Pichincha in Ecuador; all specimens studied from other departments/provinces are cited). **Colombia.** NARIÑO: between San Isidro and La Planada, Ricaurte, *Benavides 9244* (US); towards Altaquer, 43 km W of El Espino, *Bernal & Hammel 1291* (COL, MO); La Planada, *Betancur et al. 2594* (HUA); across Río San Juan from Maldonado, Ecuador, *Boeke 851* (NY, SEL); Río San Juan, San Antonio, *Ewan 16073* (AAU, US); above San Miguel, Curcuele, 8 km below Piedrancha, Río Guabo valley, *Fosberg 21066* (NY, US); trail to El Hondón, 5–12 km SW of La Planada, *Gentry et al. 60481* (MO); trail above San Juan river, near Maldonado, *Hoover 1060* (MO, QCA); Pususquer, Tuquerres, *Lehmann 5846* (F, US); La Planada, E of Tumaco, *León et al. 1676* (US); 2 km from Chambúen on the Ricaurte–Junín rd, *Londoño et al. 494* (US); Mpio. Ricaurte, La Planada reserve, *Stiles 563* (COL-2 sheets). **Ecuador**. CARCHI: Maldonado, 90 km W of Tulcan, *Balslev 1954* (QCA, QCNE, US); Quebrado Naranjo, Maldonado, *Dorr & Barnett 6090* (NY, QCA, QCNE); Chical, W of Maldonado, *Gentry & Shupp 26391* (GB, MO, QCA, SEL, US); Chical, Río Pablo, *Hoover & Wormley 1424* (MO); Río Verde, *Hoover & Wormley 1526* (MO); Untal, *Hoover & Wormley 1630* (MO, QCA); Maldonado, *Madison et al. 4440* (SEL); Chical, *Madison et al. 4454* (SEL), *4658* (SEL); Maldonado, *Werling & Leth-Nissen 393* (AAU, QCA).—COTOPAXI: Cantón La Mana, rd between Guayacán and Montenuevo, near Escuela Quindigua, *Croat 73736* (MO, QCNE).—ESMERALDAS: Lita–San Lorenzo rd, *Gentry et al. 69962* (US); rd Esmeraldas–Atacamas, *Paredes 40* (QCA).—IMBABURA: Cotocachi, Hda. La Florida, *Alvarez & Castro 602* (US).—PICHINCHA: old Quito–Santo Domingo rd, Las Palmeras, *Luteyn & Berg 14366* (MO, NY, QCA, US); trek from Lloa to Mindo, S side of Río Cinto, *Clark 4495* (AAU, COL, MO, QCA, QCNE, US); Cordillera de los Yumbos, *Clark 4591* (MO, QCA, QCNE, US); old Quito–Santo Domingo rd, between new rd and Chiriboga, *Mendoza-T. et al. 620* (AAU, COL, QCA, QCNE, US); near confluence of Río Guaycuyacu and Río Guayabamba, *Mendoza-T. et al. 536* (AAU, QCA, QCNE, US).

Gasteranthus quitensis has dense inflorescences and small urceolate corollas (maximally 2.5 cm long) (Fig. 4D); however, vegetative characters distinguish the species more easily. The leaves have blades with many secondary veins (13–18 per side), maximally with 7 mm distance between the veins (and usually 5 mm or less), and the blades are usually somewhat rugose between the veins on the upper surface. In addition, the venation of *G. quitensis* differs from that of other species that occasionally may have equally many and close secondary veins by having the indumentum of the upper leaf surfaces located in bands between the secondary veins and by being uninterrupted by conspicuous intersecondary veins.

Most of the collections of *G. quitensis* come from the province of Pichincha in Ecuador, probably reflecting more collecting activity in Pichincha than elsewhere, although *G. quitensis* is in fact the most common species of *Gasteranthus* in Pichincha (pers. obs.). The collection from the Ecuadorian province of Esmeraldas came from near sea level (Fig. 35). This occurrence may reflect a casual dispersal rather than the existence of significant populations in Esmeraldas.

29. Gasteranthus recurvatus L. E. Skog & L. P. Kvist, Brittonia 46: 327, fig. 5. 1994.—TYPE: COLOMBIA. Valle del Cauca: trail from Campoalegro into area controlled by Corporación Valle del Cauca, 400–500 m, 16 Feb 1989, *Smith et al. 1314* (holotype: COL!; isotypes: SEL! TULV! US! WIS!).

Herbs to subshrubs; stems erect, 1–2 m tall, to 8 mm in diameter, glabrous to arachnoid distally, internodes 1.5–4.5 cm long, quadrangular. Leaves opposite, isophyllous; blades 10–28 cm long, 5–12 cm wide, ovate, elliptic to obovate, base attenuate to acute, oblique, apex acute to acuminate, margin serrate, flat, adaxially dark green, glabrous but puberulent as juvenile, abaxially lighter green, puberulent with puberulent to tomentose veins, lateral veins 8–10 per side, stomatal clusters not conspicuous; petioles 1–4 cm long, puberulent-papillose. Inflorescences open cymes of 1–6 flowers, puberulent-tomentose; peduncles 5–12 cm long; pedicels 1–4 cm long. Calyx dark green, outside glabrous to sparsely puberulent, inside glandular-puberulent, lobes basally connate for 2–5 mm, subequal to unequal, each of the lateral lobes and the dorsal lobe ovate, dorsal lobe keeled; ventral lobes lanceolate to ovate, apex acute to obtuse, dorsal lobe ca. 15 mm long, 12 mm wide, ventral lobes ca. 10 mm long, 8 mm wide, lateral lobes ca. 12 mm long, 12 mm wide, denticulate to subentire, dorsal lobe directed backward below spur or upward, ventral lobes directed forward, lateral lobes directed upward. Corolla 3–3.5 cm long, urceolate with a small limb and a ventral pouch, sometimes ventricose, tube 2–2.5 cm long, widened from 0.7–1.0 cm at base to maximally 0.9–1.3 cm in diameter at widest point, pink-red, outside glandular-pilose particularly dorsally and distally, inside glabrous, markings not seen; spur 8–10 mm long, apex blunt, directed upward; throat 3–4 mm in diameter; limb red, lobes 1–2 mm long, 2–3 mm wide, equal, rotund, glabrous. Filaments ca. 10–11 mm long, briefly adnate to corolla tube base, glabrous; anthers ca. 2.5 mm long, 1.5 mm wide. Nectary thickened dorsally, annular, 2–3 mm high, tomentose distally, glabrescent basally. Ovary ca. 4 mm long, 4 mm in diameter, glabrous-puberulent; style 10–12 mm long, glabrous; stigma stomatomorphic to weakly bilobed. Capsule ca. 8 mm long, ca. 10 mm wide, flattened, bivalved, dehiscing secondarily into 4 valves; seeds ca. 0.2 mm long, 0.1 mm wide, subellipsoid, somewhat striate, with surface crystals. Figs. 10C, 36.

Phenology. Collected in flower from January to March, May to July, and in September; in fruit during February.

Distribution (Fig. 19). Colombia (Valle del Cauca) and Ecuador (Carchi); on steep wet forested slopes; 400–1600 (–1900) m.

ADDITIONAL SPECIMENS EXAMINED. **Colombia**. VALLE DEL CAUCA: Buga–Loboguerrero rd along Calima lake, *Croat 56733* (MO); Río Dagua valley, Río San Juan above Queremal, Km 52, *Cuatrecasas 23978* (F-2 sheets, US); Mpio. Dagua, Correg. el Danubio, Alto Anchicayá, *Devía 594* (MO, US); NW of Darién, Río Bravo, *Robinson 81* (COL, K, US); Calima Gorge, *Robinson 391* (COL, K, US); Anchicayá, *Vogel 93* (US). **Ecuador**. CARCHI: Reserva Etnica Awá, Community Gualpi Medio along Río Canumbí, *Grijalva et al. 604* (MO, QCNE).

Gasteranthus recurvatus (Figs. 10C, 36) differs from all other species of *Gasteranthus* with urceolate corollas by having a spur that is bent upward rather than pointing more or less straight backward or being bent downward. Only one other species, *G. imbricans* (Fig. 8C) from Panama and Costa Rica, has an upward-pointed spur, but that species has much smaller ventricose corollas (less than 2.2 cm long rather than more than 3 cm). Partly reflecting the upward bent spur, the corolla tube of *G. recurvatus* is also considerably wider at the base (from 7 to 10 mm) than that found in other species with similar-sized corollas. The flowers of *G. recurvatus* are also unusual by having corollas that are glandular-pilose outside, combined with calyx lobes that are glandular pubescent inside. The inflorescences, with pedicels up to 4 cm long, are more open than in most species of *Gasteranthus*.

The single Ecuadorian collection is similar to those from Colombia but lacks corollas, and can therefore be referred only tentatively to *G. recurvatus*; it is not included on the distribution map (Fig. 19).

FIG. 36. *Gasteranthus recurvatus.* A. Habit. B. Inset: section of abaxial leaf surface. C. Corolla. D. Corolla opened to show stamens. E. Calyx. F. Calyx, pistil, and disk. G. Fruit. H. Seeds. (Based on: A–C, E, F, *Robinson 391*; D, *Devia 594*; G, H, *Vogel 93*.)

30. Gasteranthus tenellus L. E. Skog & L. P. Kvist, sp. nov.—TYPE: ECUADOR. Esmeraldas: Quinindé, Community Piedrita, 10 km SW of Cube (via Pircuta), 400–750 m, 00°20′N, 79°41′W, 11 Jul 1996, *Clark et al. 2815* (holotype: QCNE!; isotypes: AAU! COL! E! MO! NY! QCA! SRP! US! VEN!).

A Gasterantho delphinioide (Seemann) Wiehler inflorescentiis epeduncularibus unifloris, in caulibus glabris, et foliis primariis usque ad 12 cm longis differt.

Herbs; stems decumbent to erect, to 50 cm tall, to 4 mm in diameter, glabrous, internodes 1–4 cm long, terete. Leaves opposite, isophyllous; blades 8–13 cm long, 3–6 cm wide, ovate to broadly elliptic, base cuneate, apex acute to acuminate, margin denticulate to serrate, flat, adaxially dark green, glabrous, abaxially green or reddish, reddish purple or dark reddish, glabrous, except sparsely pubescent along the veins, lateral veins mostly 6–8, stomatal clusters conspicuous; petioles 1–3 cm long, glabrous. Inflorescences open cymes of 1 flower, glabrous; peduncles absent; pedicels 1–1.5 cm long. Calyx green, glabrous, lobes free nearly to base, unequal, each of the lobes broadly ovate with acute apex, 7–20 mm long, 6–15 mm wide, denticulate, dorsal lobe directed backward and around spur, ventral lobes directed forward, lateral lobes directed upward. Corolla ca. 5 cm long, funnelform with large limb, tube ca. 2.5 cm long, not bent downward, widened from 4–8 mm at base to maximally 15–20 mm, white to yellow, outside glabrous, inside puberulent, throat usually with red-purple lines or bands, but sometimes completely yellow; spur 15–20 mm long, pointed and bent downward, 4–6 mm wide; throat 18–27 mm in diameter at widest point; limb yellow, lobes 6–10 mm long, 10–15 mm wide, spreading, subequal, rotund, apex emarginate, glabrous. Filaments 15–21 mm long, adnate to base of corolla tube for 3–5 mm, nearly glabrous; anthers ca. 2 mm long, 2 mm wide. Nectary dorsal, bilobed, ca. 2 mm high, tomentose. Ovary ca. 6 mm long, 5 mm wide, puberulent; style 12–14 mm long, puberulent at base; stigma stomatomorphic to weakly bilobed. Capsule ca. 5 mm long, 13 mm wide, flattened, splitting into 2 or 4 valves; seeds ca. 0.5 mm long, 0.2–0.3 mm wide, oblong, papillate. Figs. 3C, 37.

Phenology. Collected in flower in May, July, September, and October; in fruit during April, September, and October.

Distribution (Fig. 24). Ecuador (Esmeraldas, Manabí); along streams in low-elevation cloud forest understory; 250–800 m.

ADDITIONAL SPECIMENS EXAMINED. **Ecuador.** ESMERALDAS: Mache mountains, 15 km W of Quinindé, Bilsa Biological Reserve, 3 km W of Santa Isabela, *Abbott 15271* (QCNE); 5 km W of Santa Isabela, *Clark et al. 127* (QCNE, US); Quinindé, Community Piedrita, 10 km SW of Cube (via Pircuta), *Clark et al. 2816* (US); Quinindé, Bilsa Biological Station, Mache Mountains, 35 km W of Quinindé, 5 km W of Santa Isabel, *Clark et al. 2881* (E, MO, QCNE, US), *Clark 5573* (AAU, COL, MO, QCNE, US), *Clark & Pallis 5574* (QCNE, US); Quinindé Cantón, Mache-Chindul Ecological Reserve, Bilsa Biological Station, Mache Mountains, along ridge of La Loma de Los Guerrilleros, *Clark 2921* (AAU, E, F, MO, QCNE, US), *Clark 5573* (AAU, COL, E, MO, QCA, QCNE, SRP, US); Montañas de Mache, 35 km W of Quinindé, Bilsa Biological Station, N & NE border cut in NE part, then to Cube River tributary, then Invader Trail to Dogala Trail, *Mendoza-T. et al. 586* (AAU, QCA, QCNE, US).—MANABÍ: Pedernales Cantón, Reserva Ecológica Mache-Chindul, Comunidad Ambache (via marginal de la costa-Chindul), *Clark et al. 4204* (US).

Gasteranthus tenellus is only known from the Mache-Chindul Reserve in the Cordillera de Mache in Esmeraldas and Manabí provinces of northwestern Ecuador; the species is apparently relatively uncommon (and the rarest of the five sympatric *Gasteranthus* taxa found in the Bilsa cloud forest).

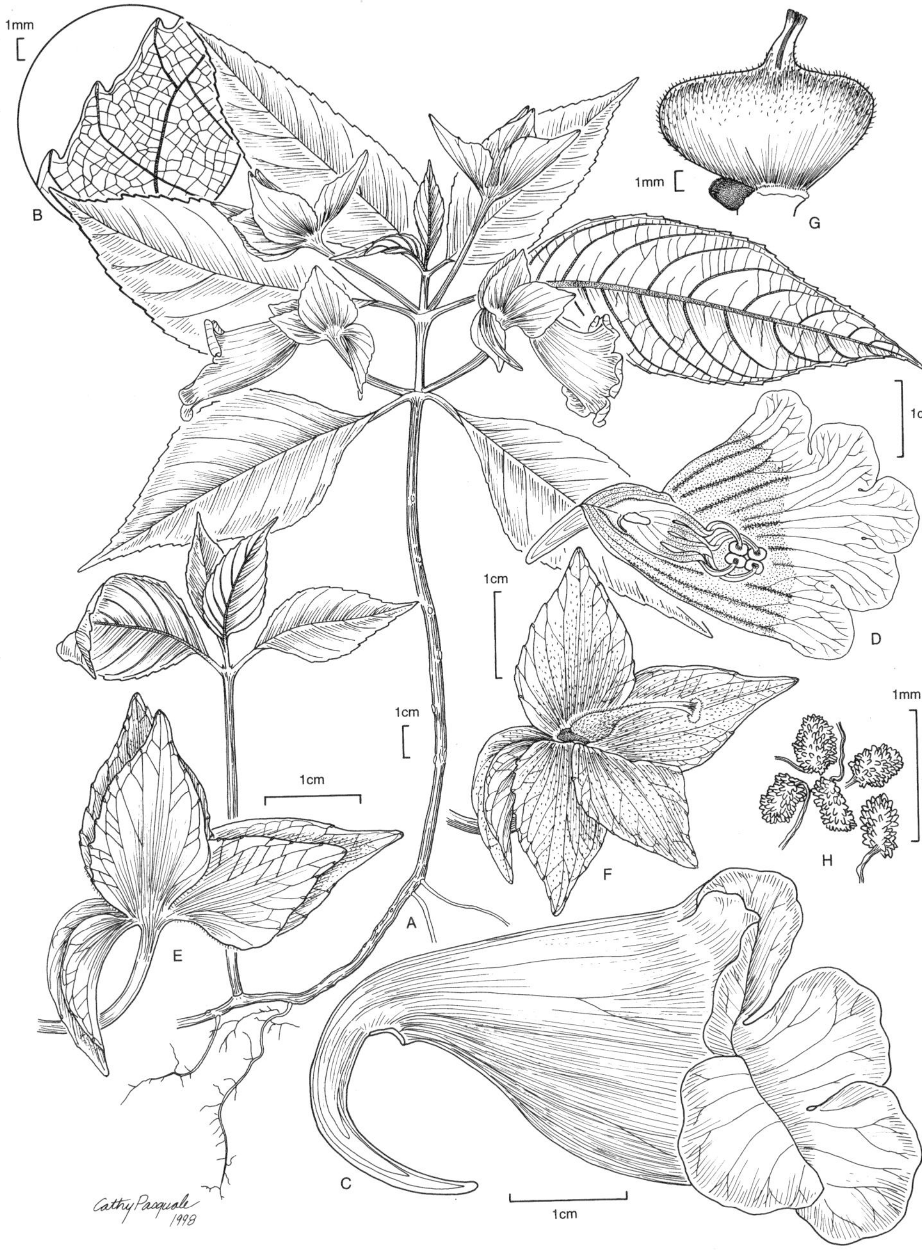

FIG. 37. *Gasteranthus tenellus*. A. Habit. B. Inset: section of abaxial leaf surface. C. Corolla. D. Corolla opened to show stamens. E. Calyx. F. Calyx, pistil, and disk. G. Fruit. H. Seeds. (Based on: A, *Abbott 15271*; B–F, *Clark et al. 2815*; G, H, *Mendoza-T. et al. 586*.)

Gasteranthus tenellus resembles *G. delphinioides*, but *G. tenellus* differs primarily in the plants being more slender and having epedunculate inflorescences; the corollas differ in shape (Fig. 3C). Initially the authors, having seen populations of *G. delphinioides* in the field in Panama and Colombia and of *G. tenellus* in Ecuador, thought their similar morphology indicated that the species were very closely related. In fact, a single collection from Tumaco nearby in the Colombian department of Nariño has features intermediate between the Bilsa population of *G. tenellus* in Ecuador and typical *G. delphinioides* populations further to the north. The Tumaco collection has corollas with a shape similar to that of *G. tenellus*, but pedunculate inflorescences and a conspicuous indumentum similar to those in *G. delphinioides*, which suggests that the Bilsa populations may not be sufficiently distinct to merit specific recognition. In this work, however, epedunculate vs. pedunculate inflorescences is considered an important taxonomic character, e.g., helping to distinguish *G. dressleri* and *G. herbaceus*. Consequently, the Bilsa population is referred to *G. tenellus*, and the Tumaco collection to *G. delphinioides*.

31. Gasteranthus ternatus L. E. Skog & L. P. Kvist, sp. nov.—ECUADOR. Pichincha: Old Quito–Santo Domingo road, above Chiriboga, steep slopes in wet ravines, 1900 m, 28 Oct 1995, *Mendoza-T. et al. 623* (holotype: QCNE!; isotypes: AAU! B! COL! GB! K! MO! NY! QCA! SEL! US!).

A Gasterantho trifoliato M. Freiberg laminis ellipticis obovatis vel oblanceolatis, lobis calycum glabris anguste ovatis vel ellipticis, et nectario bilobato differt.

Herbs to subshrubs; stems erect, somewhat scandent, rising to 2 m tall, to 6 mm in diameter, strigillose, internodes 1–5 cm long, mostly terete, sometimes quadrangular, cortex wrinkled. Leaves mostly ternate but also with opposite leaf pairs and nodes with four leaves, anisophyllous; blades 3–13 cm long, 1–5 cm wide, elliptic to obovate or oblanceolate, base cuneate, apex acute to acuminate, margin serrate, flat, adaxially dark green, glabrous or sparsely pilose, abaxially often with purplish sheen, sparsely pilose to villous on the veins, lateral veins 8–10 per side, stomatal clusters lighter in color; petioles 0.5–2.5 cm long, strigillose. Inflorescences open cymes of 1–2 (–4) flowers, glabrous; peduncles 2–7 cm long; pedicels 0.3–1.5 cm long. Calyx light green with darker green veins as well as lighter dots, glabrous, lobes free nearly to base, subequal, each of the lobes ovate to elliptic with relatively narrowly acute apex, dorsal lobe keeled, 12–22 mm long, 5–12 mm wide, subentire, dorsal lobe directed forward or backward, ventral lobes and lateral lobes directed forward. Corolla 2.6–4.2 cm long, urceolate with a small limb and a ventral pouch, tube (defined as distance from base of the corolla to throat) 1.8–3.2 cm long but distance to tip of pouch 2.2–3.8 cm, widened from 4–8 mm at base to maximally 1.2–2.2 cm (with pouch) in diameter at widest point, red, outside arachnoid, sometimes pilose particularly before anthesis, inside glabrous, markings not seen; spur 2–4 mm long, blunt, straight; throat 6–9 (–12) mm in diameter; limb dark red, lobes 1–3 mm long, 3–5 mm wide, subequal-unequal, rotund, glabrous. Filaments 8–15 mm long, adnate to base of corolla tube for 1–3 mm, glabrous; anthers ca. 3 mm long, 2.5 mm wide. Nectary dorsal, bilobed, 1–1.5 mm high, glabrous. Ovary ca. 7 mm long, 5 mm in diameter, glabrous; style ca. 7 mm long, glabrous; stigma weakly bilobed. Capsule ca. 8 mm long, 12–15 mm wide, flattened, bivalved; seeds ca. 0.3 mm long, 0.2 mm wide, elliptic, striate. Figs. 5C, 11A.

Phenology. Collected in flower in April, and from August to December; in fruit during September and October.

Distribution (Fig. 23). Ecuador (Pichincha); on steep shaded wet banks in cloud forests; (1300–) 1700–2100 m.

ADDITIONAL SPECIMENS EXAMINED. **Ecuador**. PICHINCHA: Saloya, *Acosta Solís 5788* (F); old Quito–Santo Domingo rd, Las Palmeras, Km 59, *Borchsenius 91443* (AAU, QCA); old Quito–Santo Domingo rd, Km 63, *Dodson 11833* (MO, QCNE, SEL); Km 68, *Dodson et al. 10561* (SEL); Km 82, Chiriboga, *Dodson & Thurston 14248* (MO); Km 59.5 between Chiriboga and Empalme, *Jaramillo 5968* (QCA-2 sheets, US-2 sheets); Km 59, Floristic-Ecological Reserve "Río Guajalito," *Jaramillo & Zak 7997* (MO, NY, US); new Santo Domingo–Quito rd, *Kvist & Barfod 49078* (AAU); above Mindo, *Luer et al. 4743* (SEL); Km 59, old Quito–Santo Domingo rd, Las Palmeras, *Luteyn & Berg 14359* (AAU, GB, MO, NY, QCA); Nono–Puerto Quito rd, Km 59, *Luteyn & Borchsenius 13340* (QCNE); old Quito–Santo Domingo rd, Km 72–74, *Luteyn & Lebrón-Luteyn 5645* (NY); Bancos (?), NE slopes of Pichincha, *Padilla 144* (AAU).

Gasteranthus ternatus is easily distinguished by having mostly ternate or whorled rather than opposite leaves at the nodes. The other species of *Gasteranthus* that usually have ternate leaves are *G. perennis* and *G. trifoliatus*. *Gasteranthus perennis* often has much narrower blades (more than 3.5 times as long as wide), and also differs by having calyx lobes with a long acuminate apex (Fig. 4C) rather than an acute apex (Fig. 5C). *Gasteranthus trifoliatus* from Imbabura has lanceolate to oblanceolate leaves to 1.5–2 cm wide, narrower strigillose calyx lobes, and a smaller corolla (Fig. 7A). Rarely, *G. calcaratus* may also have ternate leaves, but that species has inflorescences with more flowers than *G. ternatus*. In addition, the leaves of *G. ternatus* usually are much smaller (rarely more than 10 cm long and 5 cm wide, maximally 13 cm long) than those of most other species of *Gasteranthus*, and have conspicuous stomatal clusters on the lower surfaces. *Gasteranthus ternatus* is also set apart by having pedunculate inflorescences with only 1 or 2 flowers. It also differs in its calyx lobes, which are 1.2–2.2 cm long, ovate-elliptic, glabrous, and have white blotches, and its corollas, which are 2.6–4.2 cm long, urceolate, and bear a mostly arachnoid indumentum outside (Fig. 5C).

Specimens of *G. ternatus* had mostly been confused with *G. pansamalanus* (Fig. 5D), but this latter species always has single-flowered (but pedunculate) inflorescences, orange (rarely red) corollas, and shorter petioles (mostly shorter rather than longer than 1 cm).

32. Gasteranthus timidus (C. V. Morton) Wiehler, Selbyana 1: 155. 1975. *Besleria timida* C. V. Morton, Proc. Biol. Soc. Wash. 48: 56. 1935.—TYPE: ECUADOR. Pichincha: Corazón (?), 20 Jun 1876, *André K1497* (holotype: K!).

Herbs to subshrubs; stems erect, rising to 1.5 m tall, to 4 mm in diameter, glabrescent to pilose or villous distally and particularly at nodes, internodes 0.5–4 cm long, quadrangular. Leaves opposite, subisophyllous; blades 5–14 (–25) cm long, 2–6 cm wide, ovate to elliptic or oblanceolate, base attenuate to acute and decurrent, apex acute to acuminate, margin serrate, flat, adaxially green, glabrous to pilose particularly along primary vein and edge, abaxially lighter green, pilose to villous on the veins, lateral veins 9–12 per side, stomatal clusters not conspicuous; petioles 1–5 cm long, upper side pilose or villous, band of trichomes stretching from midvein of blade to node where bands from opposite leaves fuse. Inflorescences relatively congested cymes of 6–16 flowers, sparsely pilose to glabrous; peduncles 4–9 cm long; pedicels to 5 mm long. Calyx orange, outside densely glandular villous particularly distally, inside glabrous basally and puberulent to glandular-pubescent distally, lobes basally connate up to 1 mm, subequal, each of the lobes ovate with acute apex, dorsal lobe keeled, 1.5–3.5 mm long, 1.5–2.5 mm wide, erose, glandular-hairy, dorsal lobe directed backward, appressed below spur, ventral lobes directed forward, lateral lobes directed upward. Corolla 1.1–1.7 cm long, ventricose to urceolate with a small limb and a ventral pouch, tube 0.8–1.2 cm long, widened from 1–3 mm at base to maximally 0.6–1.0 cm in diameter at widest point (with pouch), orange to red, outside villous, inside glabrous, markings not seen; spur 2–3 mm long, straight, blunt; throat 4–7 mm in diameter; limb red, lobes 1–1.5 mm long, 2–3 mm wide, subequal, rotund-ovate, glabrous. Filaments 6–8 mm long, adnate to base of corolla tube for 1–2 mm, glabrous; anthers ca. 1 mm long, 1 mm wide. Nectary dorsal, bilobed to semi-annular and thickened, to 1 mm high, very sparsely pilose. Ovary ca. 3 mm long, 3 mm in diameter, sparsely puberulent; style 3–5 mm long, glabrous; stigma stomatomorphic. Capsule size and shape unknown, bivalved, partly splitting secondarily into 4 valves; seeds ca. 0.2 mm long, 0.1–0.2 mm wide, elliptic, papillate. Fig. 8D.

Phenology. Collected in flower in January, March to May, and July; in fruit during April and July.

Distribution (Fig. 20). Ecuador (Bolívar, Los Ríos, Manabí, Pichincha); in dense low-elevation cloud forests; 80–600 m.

ADDITIONAL SPECIMENS EXAMINED. **Ecuador**. BOLÍVAR: Balsapamba, *André 4076* (K).—LOS RÍOS: Río Pita, Hacienda Clementina, *Asplund 5315* (F); rd from Patricia Pilar to 24 de Mayo, Km 12, El Centinela at Montañas de Ila, *Dodson & Gentry 10284* (MO); 14 km W of Buena Fé, Hacienda Manfrei, *Gentry & Loor C. 30778* (MO, SEL); Centinela Ridge area, 12 km E of Patricia Pilar, *Hansen et al. 7816* (MO, SEL), *7826* (SEL); Hacienda Clementina, Cerro Sumama, trail from Destacamento Pita to La Torre, *Ståhl et al. 2119* (AAU).—MANABÍ: between Nueve de Octubre and Las Brisas, upper Río Jama valley, San Isidro area, *Pearsall 1086* (AAU); Chone–Santo Domingo rd, near Río La Morena 15 km NNE of Flavio Alfaro, *Harling & Andersson 18907* (F, GB, NY, US).

The small ventricose to urceolate corollas (maximally 1.7 cm long) and calyx lobes (maximally 3.5 mm long), both with a conspicuous glandular-villous indumentum (Fig. 8D), distinguish *G. timidus* from all other species. *Gasteranthus imbaburensis* also has the calyx and corolla with glandular trichomes, but this species has larger corollas and in particular a much larger calyx, which partly hides the corolla. The most similar corollas are probably found in *G. imbricans* from Panama and Costa Rica, but this species has much larger, imbricate calyx lobes and nearly glabrous corollas (Fig. 8C). The vegetatively most similar species is *G. calcaratus*, and *G. timidus* is sympatric with *G. calcaratus* subsp. *oncogastrus*. *Gasteranthus calcaratus* differs by having petioles with a pilose indumentum localized on the upper side, and by usually having a band of trichomes across the node between the bases of the two opposing leaves. In addition, *G. timidus* does not, in contrast to *G. calcaratus* subsp. *oncogastrus*, have conspicuous stomatal clusters on the lower leaf surfaces.

In contrast to the other low-elevation cloud forest species found in Los Ríos near the border with Pichincha province on the Centinela ridge (*Gasteranthus atratus*, *G. carinatus*, *G. extinctus*, *G. macrocalyx*, and *G. perennis*), *G. timidus* had a wider distribution, including seasonal rainforests below 300 m elevation, and possibly also in the cloud forests in the coastal hills of Manabí province. Today all these species are probably close to extinction. André's collections from the provinces of Bolívar and Pichincha have not been supplemented by recent collections.

33. Gasteranthus trifoliatus M. Freiberg, Phyton (Horn) 36: 308, fig. 2B. 1996.—TYPE: ECUADOR. Imbabura: Los Cedros Biological Reserve, 1000–1600 m, 17 Mar 1996, *Freiberg 96006* (holotype: QCA!, photo: US!; isotypes: QCNE, ULM!).

Shrubs; stems erect, rising to 1.5 m tall, 4 mm in diameter but likely larger below, strigillose, internodes 3–5 cm long, mostly terete, pilose. Leaves ternate, isophyllous; blades 10–12 cm long, 1.5 (–2) cm wide, lanceolate to oblanceolate, base attenuate, apex acute to acuminate, margin crenulate to subentire, flat, adaxially dark green, glabrous, abaxially olive-green, pilose on the veins, lateral veins 9–11 per side, stomatal clusters not conspicuous; petioles 0.5–1.0 cm long, pilose. Inflorescences reduced cymes of 1–2 flowers, sparsely puberulent; peduncles 3–6 cm long; pedicels 0.5–1 cm long. Calyx green to reddish, sparsely strigillose, lobes basally connate 1–2 mm, subequal, each of the lobes ovate with relatively broadly acute apex, 12–15 mm long, 4–6 mm wide, entire, dorsal lobe directed backward, ventral lobes directed forward, lateral lobes directed upward. Corolla 2.0–2.2 cm long, urceolate with a very small limb and a ventral pouch, tube (defined as distance from base of the corolla to throat) 1.3–1.5 cm long but distance to tip of pouch ca. 3.5 cm, widened from 5–6 mm at base to maximally 1.2 cm (with pouch) in diameter at widest point, "reddening" (according to Freiberg), outside villous, inside not seen, markings not seen; spur 2–3 mm long, blunt, straight; throat ca. 3 mm in diameter; limb red, lobes ca. 2 mm long, 3 mm wide, equal, rotund, glandular. Filaments ca. 12–14 mm long, adnate to base of corolla tube for ca. 7 mm, glabrous; anthers ca. 1 mm long, 2 mm wide. Nectary dorsal, semi-annular, ca. 1 mm high, indument unknown. Ovary ca. 5 mm long, 2 mm in diameter, glabrous; style ca. 6 mm long, glabrous; stigma bilobed. Capsule 6–7 mm long, 6–7 mm wide, globose, bivalved; seeds ca. 0.4 mm long, 0.2 mm wide, elliptic, tuberculate. Fig. 7A.

Phenology. Collected in flower and fruit only in March.

Distribution (Fig. 23). Ecuador (Imbabura); in primary premontane wet forests; 1000–1600 m.

Gasteranthus trifoliatus is very similar to *G. ternatus*, also from montane cloud forests but from Pichincha province, south of the Imbabura locality of *G. trifoliatus*. Additional collections from intervening areas may possibly demonstrate that the two species represent extremes of the same variable species. Both species usually have ternate leaves, but the narrower leaves of *G. trifoliatus* are lanceolate to oblanceolate compared to the mostly elliptic to obovate leaves of *G. ternatus* (see discussion above under *G. ternatus*, no. 31). *Gasteranthus trifoliatus* also differs by having smaller corollas than *G. ternatus*, but more notably in the calyx lobes. *Gasteranthus trifoliatus* (Fig. 7A) has narrower and acute calyx lobes (12–15 mm long, 4–6 mm wide) and is uniformly green to reddish outside, whereas *G. ternatus* (Fig. 5C) has longer and broader lobes (12–22 mm long, 5–12 mm wide), which are always light green with darker green longitudinal veins as well as lighter blotches, particularly towards the apex.

34. Gasteranthus villosus L. E. Skog & L. P. Kvist, sp. nov.—TYPE: ECUADOR. Esmeraldas: 35 km W of Quinindé, Mache Mountains, Bilsa Biological Station, 400–600 m, 26 Oct 1995, *Mendoza-T. et al. 594* (holotype: QCNE!; isotypes: AAU! COL! E! GH! MO! NY! P! QCA! US!).

A Gasterantho acropodo (J. D. Smith) Wiehler caulibus inflorescentiisque conspicue pilosis vel villosis, foliis plerumque oblanceolatis, lobis calycum pilosis, et trichomatibus eglandulosis differt.

Herbs; stems erect, to 1 m tall, to 5 mm in diameter, pilose to villous distally, internodes 1–4 cm long, terete. Leaves opposite, isophyllous; blades 7–14 cm long, 2–6 cm wide, oblanceolate, sometimes obovate or elliptic, base cuneate, apex acute to acuminate, margin serrate, flat, adaxially green, pilose, abaxially lighter green, pilose, particularly on the veins, lateral veins 8–11 per side, stomatal clusters not conspicuous; petioles 1–3 cm long, pilose-villous. Inflorescences open cymes of 1–2 (–3) flowers, pilose; peduncles 2–6 cm long; pedicels 0.5–3 cm long. Calyx green, pilose, trichomes eglandular, lobes free nearly to base, unequal, each of the lobes ovate to lanceolate with acuminate apex, 10–15 mm long, 8–12 mm wide, serrate to erose, dorsal lobe directed backward, appressed to spur, ventral lobes directed forward, lateral lobes directed upward. Corolla 3.2–4.5 cm long, ventricose with relatively small limb, tube 2.2–2.8 cm, widened from 3–5 mm at base to medially 13–19 mm in diameter at widest point, cream to yellow, outside pilose, inside puberulent basally, marked with purple-brown longitudinal stripes; spur 5–8 mm long, 4–7 mm wide, blunt, directed downward; throat 11–16 mm in diameter; limb cream or yellow, lobes 4–7 mm long, 6–8 mm wide, subequal, rotund, glabrous. Filaments ca. 10 mm long, adnate to base of corolla tube for ca. 3 mm, glabrous; anthers ca. 2 mm long, 1 mm wide. Nectary dorsal, bilobed, 2.5 mm high, densely pubescent. Ovary 3–5 mm long, 2–4 mm in diameter, puberulent; style ca. 6 mm long, puberulent; stigma stomatomorphic. Capsule ca. 3 mm long, ca. 9 mm wide, flattened, bivalved, partly splitting into 4 valves; seeds ca. 0.2 mm long, 0.1 mm wide, elliptic, papillate. Figs. 9D, 13G, H, 38.

Phenology. Collected in flower in February, March, and June to November; in fruit during October and November.

Distribution (Fig. 15). Colombia (Nariño) and Ecuador (Esmeraldas); in shady areas in mature premontane wet forests or cloud forests; (10–) 300–650 m.

ADDITIONAL SPECIMENS EXAMINED. **Colombia**. NARIÑO: Mpio. Tumaco, Santa María, *Romero-Castañeda 2736* (COL-2 sheets); Tumaco, Santa María, Río Rosario, *Romero-Castañeda 5176* (COL). **Ecuador**. ESMERALDAS: 3 km SE of San Lorenzo, *Boom 2636* (NY, QCA, US); Mache Mountains, Bilsa Biological Station, *Clark 2* (MO, QCNE, US), *Clark & Mora 337* (US), *Clark et al. 1597* (QCNE, US), *1616* (MO, QCNE, US), *Clark 5590* (MO, QCA, QCNE, US); Bilsa Biological Station, Cube River, *Clark et al. 4610* (AAU, COL, E, MO, QCA, QCNE, SRP, US); Bilsa, *Bass et al. 253* (MO, QCNE, US); San Lorenzo, *Játiva & Epling 744* (S, UC); trail to Deforno fly camp in Panadero, *Játiva & Epling 2010* (MO, NY, S, UC, US); Bilsa Biological Station, Rana Roja trail, *Mendoza-T. et al. 556* (AAU, QCA, QCNE, US); carretera Herrera–El Paramó (Sta. Isabel), Estación Biológica Bilsa, *Palacios et al. 13514* (MO).

The species most similar to *Gasteranthus villosus* (Fig. 9D) is *G. acropodus* (Fig. 9C) from Panama and Costa Rica. The two species have similar medium-sized (mostly 3–4 cm long) ventricose corollas with a relatively small limb. *Gasteranthus villosus* has conspicuously pilose to villous stems and inflorescences (Fig. 38), whereas *G. acropodus* appears glabrous, although it mostly has an appressed pubescence. In addition, *G. villosus* has conspicuously serrate, mostly oblanceolate leaf blades with more secondary veins per side (8–11) in contrast to *G. acropodus*, which has mostly ovate-elliptic, subentire leaf blades with fewer secondary veins (7–8). Also, *G. acropodus* differs by having glandular trichomes on its calyces.

35. Gasteranthus wendlandianus (Hanstein) Wiehler, Selbyana 1: 156. 1975. *Besleria wendlandiana* Hanstein, Linnaea 34: 318. 1865.—TYPE: COSTA RICA. Cartago: Coliblanco "Cori Blanco," Turrialva [Turrialba], 30 May 1857, *Wendland 922* (lectotype, designated by Morton, 1939: B, destroyed, photo: US!; lectotype, here designated: GOET!).

FIG. 38. *Gasteranthus villosus*. A. Habit. B. Inset: section of abaxial leaf surface. C. Corolla. D. Corolla opened to show stamens. E. Calyx. F. Calyx, pistil, and disk. G. Fruit. H. Seeds. (Based on *Játiva & Epling 744*).

Besleria rupestris C. V. Morton, Proc. Biol. Soc. Wash. 48: 57. 1935. *Gasteranthus rupestris* (C. V. Morton) Wiehler, Selbyana 1: 155. 1975.—TYPE: ECUADOR. Tungurahua: Pastaza river, 1300–1800 m, 25 Sep 1923, *Hitchcock 21820* (holotype: US!).

Herbs to subshrubs; stems usually erect, sometimes decumbent, rarely exceeding 1.5 m but reported up to 3 m tall, to 8 mm in diameter, lanate-tomentose when young and glabrescent with age, internodes 1–5 (–8) cm long, mostly terete, also more or less quadrangular, bark often somewhat longitudinally striate. Leaves opposite, subisophyllous; blades 7–21 cm long, 3–9 cm wide, mostly elliptic, occasionally narrowly elliptic to oblanceolate, base cuneate to acute, apex acute to acuminate, margin serrate to serrulate, flat, adaxially green to dark green, usually glabrous, rarely more or less puberulent to appressed pilose, abaxially lighter green, tomentose at veins, lateral veins 9–13 per side, stomatal clusters not conspicuous; petioles 0.5–2 cm long, mostly more or less tomentose. Inflorescences relatively congested cymes of 3–10 flowers, sparsely pilose to pilose, rarely villous; peduncles nearly perpendicular to the stem, 3–7 (–10) cm long; pedicels up to 1.2 cm but rarely exceeding 7 mm long. Calyx green to pale yellow, more or less pilose outside, inside glabrous or puberulent, lobes basally connate 1–2 mm, subequal, each of the lobes broadly to narrowly lanceolate with acuminate apex, dorsal lobe keeled, 6–9 mm long, 2–4 mm wide, longer and broader in fruit, dentate to serrate, dorsal lobe usually directed backward below spur and from there often upward, ventral lobes directed forward, lateral lobes directed forward-upward. Corolla 1.3–2.5 cm long, funnelform-tubular with relatively large limb, tube 0.7–1.5 cm, widened from 3–5 mm at base to 4–7 mm, often hardly widened at all, whitish, pale yellow to yellow, with red-purple spots, outside pilose, inside glabrous, with red or purple spots or lines in throat; spur 1–4 mm long, to 4 mm wide, blunt, straight; throat 4–7 mm in diameter; limb yellow but usually with some red or purple spots, lobes 3–4 mm long, 4–5 mm wide, subequal, rotund, glabrous. Filaments 7–13 mm long, adnate to base of corolla tube for 2–3 mm, glabrous; anthers ca. 1 mm long, 1 mm wide. Nectary thickened posteriorly, bilobed to annular and thickened dorsally, to 0.5 mm high, wooly distally and towards ovary, glandular basally and externally. Ovary 3–4 mm long, 2–4 mm in diameter, puberulent; style 4–6 mm long, puberulent to nearly glabrous; stigma stomatomorphic to weakly bilobed. Capsule 6–7 mm long, 6–7 mm wide, globose, bivalved; seeds ca. 0.3 mm long, 0.2 mm wide, elliptic, striate. Figs. 9A, 39.

Phenology. Collected in flower in all months of the year; in fruit in January to April, July, and September to December.

Distribution (Fig. 40). Costa Rica (Alajuela, Cartago, Guanacaste, Heredia, Limón, Puntarenas, San José), Panama (Bocas del Toro, Chiriquí, Coclé, Darién, Veraguas), Bolivia (Cochabamba, La Paz), Colombia (Antioquia, Cauca, Chocó, Nariño, Putumayo, Valle del Cauca), Ecuador (Morona-Santiago, Napo, Pastaza, Sucumbíos, Tungurahua, Zamora-Chinchipe), Peru (Amazonas, Ayacucho, Cuzco, Huánuco, Pasco, Puno, San Martín, Ucayalí); in densely shaded understory along streams and on slopes in montane wet forests and cloud forests; (600–) 900–1900 (–3200) m.

REPRESENTATIVE SPECIMENS. **Costa Rica.** ALAJUELA: San Ramón Forest Reserve, Río San Lorencito station, *Gómez-Laurito et al. 11848* (F); Monteverde Cloud Forest Reserve, Peñas Blancas river valley on Atlantic slope, *Haber ex Bello C. 5808* (MO).—CARTAGO: SE of Orosi, ca. 2.2 km SSE of Purisil, above Finca la Concordia, *Lellinger & White 1487* (F, US); vic. Pejivalle, *Skutch 4648* (NY, US); Turrialba, *Wendland 568* (GOET).—GUANACASTE: Río Chiquito de Tilarán, Río Negro valley, *Haber & Bello 5876* (MO).—HEREDIA: Yerba Buena, NE of San Isidro, *Standley & Valerio 49806* (US); ravine above San Miguel in valley of Río Sarapiquí, *Moore 6608* (US).—LIMÓN: P. N. Cordillera de Talamanca, junction of Quebrada Kuisa with Río Lori, between Ujarrás–San José Cabécar, *Herrera 5949* (MO).—PUNTARENAS: Monteverde Cloud Forest Reserve, *Feinsinger et al. 465* (F); Coto Brus, Amistad Biosphere Reserve, near Las Alturas de Cotón Biological Station, *Kress & Cablk 94-3845* (US).—SAN JOSÉ: La Hondura, *Standley & Valerio 51798* (US); ca. 25 km N of San

FIG. 39. *Gasteranthus wendlandianus*. A. Habit. B. Lateral view of flower. C. Face view of corolla. D. Corolla opened to show stamens, pistil, and disk. E. Calyx and fruit. F. Seeds. (Based on *Skog & Skog 5199*.)

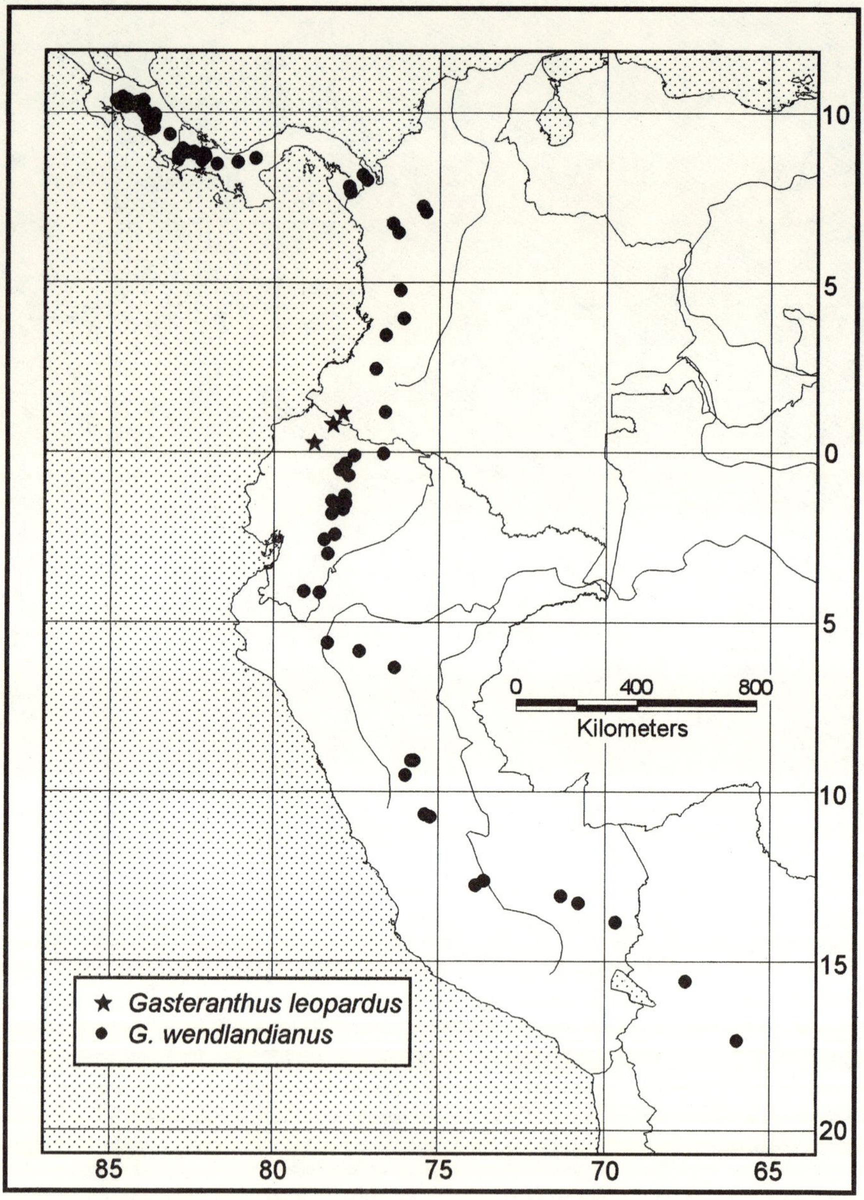

FIG. 40. Distribution of *Gasteranthus leopardus* and *G. wendlandianus*.

Isidro de El General along Pan American Highway, *Williams et al. 28588* (F, NY). **Panama.** BOCAS DEL TORO: rd from Fortuna Lake to Chiriquí Grande, 8 km N of continental divide, *Hampshire & Whitefoord 510* (BM).—CHIRIQUÍ: above Boquete on trail to Bocas del Toro along Río Caldera, *McPherson & Merello 8352* (MO); Fortuna Dam area, *D'Arcy et al. 16018* (MO, US).—COCLÉ: El Copé, *Hammel 13637* (US).—DARIÉN: Cerro Pirre, *Folsom 4367* (US).—VERAGUAS: Cerro Tuta, E slopes, 1 km beyond Alto Piedra above Santa Fé, *Sytsma & Andersson 4647* (NY, US).

Colombia. ANTIOQUIA: Mpio. Frontino, rd to Murrí, 22.7 km W of Nutibara, *Brant & Martínez A. 1434* (HUA); Mpio. Briceño, Vereda San Fermín, rd Ventanas (Mpio. Yarumal)–Briceño, Km 2.5, *Callejas et al. 5543*

(HUA).—CAUCA: rd from El Tambo to 20 de Julio, *Plowman & Vaughan 5319* (GH, SEL).—CHOCÓ: top of Serrania del Darién, E of Unguia, *Gentry et al. 16783* (MO).—NARIÑO: Mpio. Barbacoas, Chucunes–La Planada road, *Lozano C. et al. 6887* (COL).—PUTUMAYO: San Francisco de Sibundoy, trail to Río Minchoy, *Londóño & Lozano 226* (AAU).—VALLE DEL CAUCA: Mpio. La Elvira, 25 km W of Cali, Finca Zingara, *Luteyn et al. 12545* (NY, US). **Ecuador**. MORONA-SANTIAGO: between La Esperanza and Santa Ana, Huamboya, *Acosta Solís 7435* (US); Colonia Azuay, 2 km from Arapicos, *Lugo S. 6090* (F, GB, NY, QCA, US).—NAPO: Archidona Cantón, Reserva Ecológica Antisana, comunidad Shamato, entrada por Km 21–Shamato, *Clark et al. 5061* (AAU, MO, QCA, QCNE, US); Cerro Antisana, Shinguipino Forest between Río Napo and Tena, 8 km SE of Tena, *Grubb et al. 1594* (BH); El Chaco–Río Oyacachi, between Chonta Loma and Río Oyacachi, *Jaramillo et al. 12349* (QCA).—PASTAZA: Mera, 4 km N of village, *Bohlin & Bohlin 711* (GB, US); Pomona, ca. 10 km E of Puyopungu, *Lugo S. 5155* (F, GB, QCA, SEL).—SUCUMBÍOS: rd Baeza–Lago Agrio, Río Salado, 49 km from Baeza, *Balslev & Madsen 10269* (AAU, F, MO, NY, SEL); Reserva Ecológica Cayambe Cosa, Volcán Reventador, between Río Quijos and Refugio, *Clark 4456* (MO, QCNE, US).—TUNGURAHUA: valley of Pastaza River, between Baños and Cashurco, *Hitchcock 21820* (US).—ZAMORA-CHINCHIPE: rd Loja–Zamora, Km 49, *Holm-Nielsen et al. 3799* (AAU, F, MO, NY, US); Nangaritza Cantón, Río Nangaritza, Pachicutza, *Palacios et al. 8196* (QCNE). **Peru**. AMAZONAS: Prov. Bagua, Cordillera Colán, SE of La Peca, *Barbour 3841* (MO, US).—AYACUCHO: Prov. La Mar, between Tambo San Miguel, Ayna and the Hacienda Luisiana, *Dudley 11931* (F).—CUZCO: Prov. Paucartambo, Km 141 on Paucartambo–Manu rd (Peru 26B) between Pillahuata and Pilcopata, Manu National Park, *Skog & Skog 5199* (US, USM); Prov. Paucartambo, San Pedro–Sta. Isabel, *Vargas 16844* (US-2 sheets); Prov. Quispicanchi, Camanti, Maniri, towards Cerro Camanti, *Timaná 748* (MO).—HUÁNUCO: Prov. Huánuco, Dtto. Churubamba, trail Puente Durand to Exito, below Palo Marcado, *Mexia 8237* (F, GB, NY, US); Prov. Leoncio Prado, Dtto. Hermilio Valdizán, La Divisoria, *Schunke V. 11257* (IBE-2 sheets, US).—PASCO: Pichis Trail, Eneñas, *Killip & Smith 25759* (F, NY, US); Prov. Oxapampa, Abra los Mellizos, 4–8 km from Eneñas, *Skog et al. 5031* (US, USM); Prov. Oxapampa, rd between Oxapampa and Villa Rica, Km 7 SE from Miraflores crest, *Foster 9118* (MO, US).—PUNO: Prov. Sandia, Sto. Domingo area, *McCarroll 72* (MICH, NY, S).—SAN MARTÍN: Km 15, Tarapoto–Yurimaguas, *Besse et al. 782* (SEL).—UCAYALÍ: NE of pass at La Divisoria, trail off Tingo María–Pucallpa rd, *Skog et al. 5163* (US, USM); bridge over Río Chino between La Divisoria and Boquerón de Padre Abad on Tingo María–Pucallpa rd, *Skog et al. 5172* (US, USM). **Bolivia**. COCHABAMBA: NE of Cochabamba toward Villa Tunari, *Luer & Luer 3536* (SEL).—LA PAZ: Prov. Nor Yungas, Serranio de Bella Vista, 15.1 km N of Carasco bridge on Palos Blancos rd, 36 km N of Caranavi, *Solomon 14838A* (MO).

Most specimens of *G. wendlandianus* differ from those of all other species of *Gasteranthus* in these three aspects: 1) inflorescences with peduncles nearly perpendicular to the stems, particularly in the fruiting stage; 2) calyx lobes with a conspicuously toothed margin (but in some specimens the teeth are reduced); and 3) small corollas (1.3–2.5 cm long), which are funnelform with a relatively narrow but never constricted throat (4–7 mm wide), and white to yellow with darker dots (Fig. 9A). Some specimens of three other species, *G. carinatus*, *G. columbianus*, and *G. imbricans* (all Fig. 8), have similarly small funnelform corollas. They all differ from *G. wendlandianus* by their much brighter and mostly uniformly red corollas, mostly ovate calyx lobes with acute-obtuse apices (rather than lanceolate lobes with acuminate apices) and entire-subentire margins, and stems without a conspicuous indumentum.

The species most similar to *G. wendlandianus* (Fig. 39) is *G. leopardus* (Fig. 30), found in western Andean montane forests south of the Colombian range of *G. wendlandianus*. *Gasteranthus leopardus* usually has much larger corollas (mostly between 4 and 5 cm long), with a much darker usually mostly brown limb; the two species appear to intergrade in the department of Nariño (see *G. leopardus*, no. 20).

Gasteranthus wendlandianus, the most widespread species in the genus, occurs from Costa Rica through Panama, Colombia, Ecuador, Peru, and to Bolivia; however, this range is not continuous. The northern populations, from Costa Rica south to the western Colombian department of Cauca, are isolated from the southern populations, which are found on the eastern slopes of the Andes in the Ecuadorian province of Napo in the north to the

Peruvian department of Puno and the province of Cochabamba in Bolivia in the south. *Gasteranthus wendlandianus* does not (as does the other widespread species, *G. corallinus*) occur in Amazonian lowland to the east of the Andean foothills, but it ranges much further to the south than any other species of *Gasteranthus*. Collecting gaps may also contribute to the discontinuous range. Despite the wide distribution of the species, there are no major differences among populations of *G. wendlandianus* found in different parts of the range.

Excluded Names

Besleria imbricans var. *arguta* C. V. Morton, Contr. U.S. Natl. Herb. 26: 416, vii. 1939, nomen nudum [error for *Besleria imbricans* var. *uncinata* C. V. Morton; see also discussion of *Gasteranthus imbricans*, no. 18].

Gasteranthus pendulus Hanstein, Linnaea 26: 216, pl. 2, fig. 58. 1854, nomen nudum.

Gasteranthus pendulus Klotzsch ex Hanstein, Linnaea 34: 333. 1865, pro syn.

Gasteranthus pilosus Bentham, Pl. Hartweg. 233. 1846, pro syn.

ACKNOWLEDGMENTS

The authors thank the following herbaria for permission to study their collections: A, AAU, AD, B, BH, BM, C, CAS, CGE, CM, COL, CR, CUVC, DAV, DS, DUKE, E, EAP, F, FI, FLAS, G, GB, GH, GOET, GUAY, HBG, HUA, IBE, K, L, LA, LAM, LE, M, MEDEL, MEXU, MICH, MO, MPU, NA, NO, NY, OS, P, PH, PMA, QCA, QCNE, QPLS, REED, RSA, S, SEL, SCZ, SRP, TULV, U, UC, ULM, US, USM, VEN, W, WAG, WIS, WU. We also thank John Clark, Marisol Amaya Márquez, and Paulina Mendoza-Troya for assistance in the field; John Boggan for his numerous tasks in the laboratory, including assisting with computer mapping, and growing live plants; Harold Robinson in looking over the Latin diagnoses; Anne Senters for her observations; and Cathy Pasquale-Johnson and Alice Tangerini for their excellent illustrations. Figure 36 (*Gasteranthus recurvatus*, by Cathy Pasquale-Johnson) is reprinted with permission from *Brittonia*, vol. 46, no. 4, copyright 1994, The New York Botanical Garden. The second author thanks the Carlsberg Foundation for funding his visits in Colombia, Ecuador, and the U.S.A. Distribution maps were produced with MapInfo Professional™ Version 4.1, using basemaps from ADC WorldMap™ Version 2.0. All basemaps copyright 1998, American Digital Cartography, Inc.

LITERATURE CITED

Beaufort-Murphy, H. T. 1983 ["1982"]. The seed surface morphology of the Gesneriaceae utilizing the scanning electron microscope and a new system for diagnosing seed morphology. Selbyana 6: 220–422.

Bentham, G. 1846. *Plantae Hartwegianae*. London: W. Pamplin.

_____. 1876. Gesneraceae. In *Genera Plantarum* by G. Bentham and J. D. Hooker, 2: 990–1025. London: Lovell Reeve & Co.

Brandegee, T. S. 1917. Plantae Mexicanae Purpusianae VIII. Univ. California Publ. Bot. 6: 363–375.

Burtt, B. L., and H. Wiehler. 1995. Classification of the family Gesneriaceae. Gesneriana 1: 1–3.

Dodson, C. H., and A. H. Gentry. 1978. Flora of the Río Palenque Science Center, Los Ríos, Ecuador. Selbyana 4: 1–628.

_____. 1991. Biological extinction in western Ecuador. Ann. Missouri Bot. Gard.78: 273–295.

Freiberg, M. 1996. Gesneriad flora of the Los Cedros Biological Reserve, northwest Ecuador, part 1: Four new species in *Gasteranthus* (Gesneriaceae). Phyton (Horn) 36: 303–309.

_____. 1998. Two remarkable new species of *Gasteranthus* (Gesneriaceae) from central Ecuador. Phyton (Horn) 38: 167–173.

Fritsch, K. 1922. Neue *Besleria*-Arten. Repert. Spec. Nov. Regni Veg. 18: 7–13.

_____. 1934. Zur Kenntnis der Gattung *Besleria*. Notizbl. Bot. Gart. Berlin-Dahlem 11: 961–977.

Gentry, A. H. 1989. Speciation in tropical forests. In *Tropical Biology*, ed. K. Larsen and L. B. Holm-Nielsen, 113–134. London: Academic Press.

_____. 1992. Tropical forest biodiversity: Distributional patterns and their conservational significance. Oikos 63: 19–28.

Gentry, A. H., and C. H. Dodson. 1987. Diversity and biogeography of neotropical vascular epiphytes. Ann. Missouri Bot. Gard. 74: 205–233.

Hanstein, J. 1865. Die Gesneraceen des Königlichen Herbariums und der Gärten zu Berlin, nebst monographischer Uebersicht der Familie im Ganzen, II: Gattungen und Arten, 3: Die Eugesnereen, Rhytidophylleen und Beslerieen. Linnaea 34: 225–446.

Kunth, C. 1818 ["1817"]. Gesneriaceae. In *Nova genera et species plantarum*, by F. H. A. von Humboldt, A. Bonpland, and C. Kunth, 2: 392–401 [quarto ed.]; 2: 315–321 [folio ed.], plates 188–192. Paris.

Kvist, L. P. 1990. Revision of *Heppiella* (Gesneriaceae). Syst. Bot. 15: 720–735.

Kvist, L. P., and L. E. Skog. 1989. Revision of *Reldia* (Gesneriaceae). Nord. J. Bot. 8: 601–611.

_____. 1992. Revision of *Kohleria* (Gesneriaceae). Smithson. Contr. Bot. 79: 1–83.

_____. 1993. The genus *Columnea* (Gesneriaceae) in Ecuador. Allertonia 6(5): 327–400.

_____. 1996. Revision of *Pearcea* (Gesneriaceae). Smithson. Contr. Bot. 84: 1–47.

Mansfeld, R. 1936. Gesneriaceae. In "Beiträge zur Kenntnis der Vegetation und Flora von Ecuador." Biblioth. Bot. 116: 144–147.

Morton, C. V. 1935. Synopsis of the species of *Besleria* in Ecuador. Proc. Biol. Soc. Wash. 48: 55–58.

_____. 1938. Gesneriaceae. In "Flora of Costa Rica," ed. P. C. Standley. Publ. Field Mus. Nat. Hist., Bot. Ser. 18: 1137–1187.

_____. 1939. A revision of *Besleria*. Contr. U.S. Natl. Herb. 26: 395–474.

_____. 1944. Taxonomic studies of tropical American plants. Contr. U.S. Natl. Herb. 29: 1–40.

_____. 1968. The Peruvian species of *Besleria* (Gesneriaceae). Contr. U.S. Natl. Herb. 38: 125–151.

Parker, T. A., III, and J. L. Carr, eds. 1992. Status of forest remnants in the Cordillera de la Costa and adjacent areas of southwestern Ecuador. *RAP Working Papers 2*. Washington, D.C.: Conservation International.

Seemann, B. 1854. Gesneriaceae. In *Botany of Voyage of H.M.S. Herald* by B. Seemann, 184–187, 250–251. London: Lovell Reeve.

Skog, L. E., and F. de Jesús. 1997. A review of *Resia* (Gesneriaceae). BioLlania Edición Esp. 6: 515–525.

Skog, L. E., and L. P. Kvist. 1994. Novae Gesneriaceae Neotropicarum VII: Five new Gesneriaceae from northwestern South America. Brittonia 46: 317–330.

Smith, J. D. 1891. Undescribed plants from Guatemala IX. Bot. Gaz. 16: 196–200, pl. 16–18.

———. 1912. Undescribed plants from Guatemala and other Central American Republics XXXV. Bot. Gaz. 54: 238–241.

Smith, J. F. 1994. Systematics of *Columnea* section *Pentadenia* and section *Stygnanthe* (Gesneriaceae). Syst. Bot. Monogr. 44: 1–89.

Standley, P. A., and L. O. Williams. 1951. Plantae Centrali-Americanae, II. Ceiba 1: 231–255.

Weber, A. 1973. Die Struktur der paarblütigen Partialinfloreszenzen der Gesneriaceen und bestimmter Scrophulariaceen. Beitr. Biol. Pflanzen 49: 429–460.

_____. 1978. Transitions from pair-flowered to normal cymes in Gesneriaceae. Notes Roy. Bot. Gard. Edinburgh 36: 355–368.

_____. 1982. Evolution and radiation of the pair-flowered cyme in Gesneriaceae. Australian Syst. Bot. Soc. Newsl. 30: 23–47.

Wiehler, H. 1975. *Besleria* L. and the re-establishment of *Gasteranthus* Benth. (Gesneriaceae). Selbyana 1: 150–156.

_____. 1977. New genera and species of Gesneriaceae from the neotropics. Selbyana 2: 67–132, pl. 19–36.

_____. 1978. The genera *Episcia*, *Alsobia*, *Nautilocalyx*, and *Paradrymonia* (Gesneriaceae). Selbyana 5: 11–60.

_____. 1983. A synopsis of the neotropical Gesneriaceae. Selbyana 6: 1–219.

Wilson, C. L. 1974. Floral anatomy in Gesneriaceae. II. Gesnerioideae. Bot. Gaz. 135: 256–268.

APPENDIX

Distribution of species of *Gasteranthus* by country (endemics are noted with an asterisk).

BOLIVIA: *G. wendlandianus.*

COLOMBIA: *G. adenocalyx**, *G. anomalus**, *G. calcaratus* subsp. *calcaratus**, *G. calcaratus* subsp. *calceolus*, *G. calcaratus* subsp. *oncogastrus*, *G. columbianus*, *G. corallinus*, *G. delphinioides*, *G. epedunculatus**, *G. glaber*, *G. herbaceus**, *G. leopardus*, *G. pansamalanus*, *G. quitensis*, *G. recurvatus*, *G. villosus*, *G. wendlandianus.*

COSTA RICA: *G. acropodus*, *G. delphinioides*, *G. imbricans*, *G. osaensis**, *G. wendlandianus.*

ECUADOR: *G. atratus**, *G. bilsaensis**, *G. calcaratus* subsp. *calceolus*, *G. calcaratus* subsp. *oncogastrus*, *G. carinatus**, *G. columbianus*, *G. corallinus*, *G. crispus**, *G. extinctus**, *G. glaber*, *G. imbaburensis**, *G. lateralis**, *G. leopardus*, *G. macrocalyx**, *G. mutabilis**, *G. orientandinus**, *G. otongensis**, *G. pansamalanus*, *G. perennis**, *G. quitensis*, *G. recurvatus*?, *G. tenellus**, *G. ternatus**, *G. timidus**, *G. trifoliatus**, *G. villosus*, *G. wendlandianus.*

GUATEMALA: *G. pansamalanus.*

MEXICO: *G. pansamalanus.*

PANAMA: *G. acropodus*, *G. delphinioides*, *G. dressleri**, *G. imbricans*, *G. wendlandianus.*

PERU: *G. calcaratus* subsp. *calceolus*, *G. corallinus*, *G. wendlandianus.*

NUMERICAL LIST OF TAXA

1. *G. acropodus*
2. *G. adenocalyx*
3. *G. anomalus*
4. *G. atratus*
5. *G. bilsaensis*

6a. *G. calcaratus* subsp. *calcaratus*

6b. *G. calcaratus* subsp. *calceolus*

6c. *G. calcaratus* subsp. *oncogastrus*

7. *G. carinatus*
8. *G. columbianus*
9. *G. corallinus*
10. *G. crispus*
11. *G. delphinioides*
12. *G. dressleri*
13. *G. epedunculatus*
14. *G. extinctus*
15. *G. glaber*
16. *G. herbaceus*
17. *G. imbaburensis*
18. *G. imbricans*
19. *G. lateralis*
20. *G. leopardus*
21. *G. macrocalyx*
22. *G. mutabilis*
23. *G. orientandinus*
24. *G. osaensis*
25. *G. otongensis*
26. *G. pansamalanus*
27. *G. perennis*
28. *G. quitensis*
29. *G. recurvatus*
30. *G. tenellus*
31. *G. ternatus*
32. *G. timidus*
33. *G. trifoliatus*
34. *G. villosus*
35. *G. wendlandianus*

INDEX TO NUMBERED COLLECTIONS EXAMINED

The numbers in parentheses refer to the corresponding species in the text and in the Numerical List of Taxa above.

Abbott, J. R. 15206 (6c); 15271 (30).
Acevedo R., P., et al. 1333 (2); 6787 (11).
Acosta Solís, M. 5636 (28); 5770 (22); 5782 (19); 5788 (31); 5811 (15); 6949 (9); 7053 (26); 7435 (35); 7443 (35); 12220 (9); 13652 (6c).
Allen, P. H. 1829 (1); 2188 (1); 2371 (1); 4920 (35); 5533 (24); 5589 (11).
Almeda, F., & Nakai, K. 3971 (35).
Alvarado, A. 250 (9).
Alvarez, A. 684 (6c).
Alvarez, A., & Castro, R. 602 (28); 660 (15).
Alvarez, A., & Herrera, P. 707 (6c).
Alvarez, A., et al. 986 (26).
Amaya, M., & Kvist, L. P. 377 (11); 390 (11); 404 (11); 415 (11); 452 (11).
André, E. 751 (26); K1486 (9); K1487 (28); K1488 (28); K1489 (28); K1490 (28); K1491 (28); K1492 (28); K1493 (28); K1494 (26); K1495 (6c); K1497 (32); 2372 (6a); 3329 (9); 3418 (28); 3814 (26); 3834 (19); 4076 (32); 4270 (6c); 4278 (6c).
Antonio, T. 746 (18); 1231 (1); 1312 (11); 1524 (35); 1890 (35); 3095 (1); 3114 (1); 3115 (18); 3345 (1); 3360 (12); 4178 (35); 4763 (1); 4777 (12); 5105 (35).
Arbeláez S., G., et al. 15 (6a).
Asplund, E. 5315 (32); 5523 (27).
Aulestia, C., et al. 54 (9); 691 (6c); 829 (9).
Aulestia, M., & Gonti, O. 1783 (9).
Ayala, F., et al. 2466 (9).
Baker, M., et al. 6385 (9).
Balslev, H. 1954 (28); 2410 (9).
Balslev, H., & Balseca, S. 4684 (9).
Balslev, H., & Madsen, E. 10269 (35).
Balslev, H., et al. 60705 (9); 62345 (9); 62379 (6b); 62499 (9).
Barbour, P. 3841 (35).
Barclay, H. 4860 (9).
Barfod, A. 41000 (6c); 41085 (9); 41445 (9); 60705 (9).
Barfod, A., & Skov, F. 60074 (6c).
Barfod, A., et al. 48018 (6c); 48094 (9); 48126 (9).
Barringer, K., et al. 3462 (1); 3608 (1).
Bass, M. S., & Pitman, N. 197 (6c); 203 (10).
Bass, M. S., et al. 108 (6c); 253 (34).
Benavides, O. de 2884 (15); 4070 (20); 8840 (8); 9234 (20); 9244 (28); 9621 (8); 9775 (8); 10162 (8); 10225 (15); 10914 (15); 10924 (20); 11134 (20); 11203 (20).
Berg, C. C., & Akkermans, R. W. A. P. 1077 (9).
Bernal, R., & Hammel, B. 1291 (28).
Bernardello, L. 761 (26).
Berry, P., & Brako, L. 4665 (26).
Besse, L. 16 (9).
Besse, L., et al. 12 (9); 164 (28); 782 (35); 1662 (9); 2277 (26); 2320 (26).
Betancur, J., et al. 2594 (28); 4748 (8); 4885 (8).
Billiet, F., & Jadin, B. 6674 (26).
Blasca, 1640 (19); 1641 (28).
Boeke, J. D. 851 (28); 2218 (6b).
Boeke, J. D., & Loyola, H. 1020 (26).
Boeke, J. D., & McElroy, J. B. 441 (26).
Bohlin, J.-E., & Bohlin, M. 300 (9); 711 (35).
Bohlin, J.-E., et al. 1110 (28); 1122 (28); 1478 (6b); 1559 (9).
Bonifaz, C., & Cornejo, X. 3346 (9).
Boom, B. M. 2636 (34).
Borchsenius, F. 91408 (26); 91410 (28); 91443 (31).
Boysen-Larsen & Eriksen 45255 (35).
Bradford, J., et al. 45 (15).
Brandbyge, J., & Asanza C., E. 30019 (9); 30105 (9); 32786 (9).
Brandbyge, J., et al. 30292 (6b); 32464 (9); 33191 (9).
Brant, A. E., & Martínez A., G. E. 1414 (11); 1434 (35).
Breedlove, D. E. 42568 (26).
Breedlove, D. E., & Almeda, F. 47723 (26).
Breedlove, D. E., & Smith, A. R. 31633 (26).
Brenes, A. M. 4095 (35); 5740 (35); 13573 (35); 19007 (35).
Burbano, M. 201 (9).
Burger, W. C., & Antonio, T. 10964A (18); 10964 (18).
Burger, W. C., & Gentry, J. L. Jr. 8719 (35); 9015 (24).
Burger, W. C., & Liesner, R. L. 6835 (35).
Burger, W. C., & Stolze, R. G. 5481 (24).
Cabrera R., I. 6583 (6a).
Callejas, R. 288 (11).
Callejas, R., & Bornstein, A. J. 11068 (35).
Callejas, R., et al. 2816 (11); 4804 (11); 5543 (35); 5656 (11); 5739 (11); 6166 (35); 6800 (2); 6850 (35); 9535 (6a); 9762 (11).
Camp, W. H. E-1492 (23).
Carballo, G. 223 (1).
Carleton, M. A. 205 (18).
Carrasquilla, L., & Mendoza, R. 1350 (1).
Cazalet, P. C. D., & Pennington, T. D. 5012 (6c); 7599 (6b).
Cerón M., C. E. 805 (9); 1701 (6b); 3579 (9); 6041 (9); 12403 (28).
Cerón M., C. E., & Benavides, G. 6179 (6c).
Cerón M., C. E., & Cerón, M. 3082 (35); 4592 (9).

Cerón M., C. E., & Villavicencio, S. 2806 (26).
Cerón M., C. E., et al. 2109 (9); 5913 (28); 7912 (15); 7913 (19); 7973 (26); 8901 (19).
Chacón, I. A: 257 (35).
Chavarria D., M., & Mendez, M., et al. 208 (35).
Churchill, H. W., & Churchill, A. 6232 (1).
Churchill, H. W., et al. 4718 (35).
Clark, J. L. 2 (34); 1707 (6c); 1746 (8); 2152 (6c); 2175 (6c); 2896 (6c); 2921 (30); 2964 (6c); 3687 (6c); 4456 (35); 4474 (9); 4494 (19); 4495 (28); 4694 (6c); 4695 (6c); 4696 (6c); 4712 (6c).
Clark, J. L., & Adnepos, B. 58 (10).
Clark, J. L., & Brame, C. 1702 (6c); 1715 (6c).
Clark, J. L., & Clark, S. 3910 (28).
Clark, J. L., & Dunn, R. W. 2435 (26); 2440 (9); 2443 (26); 2444 (20); 2450 (8); 2462 (15); 2463 (20); 2467 (15).
Clark, J. L., & Mora, S. 337 (34).
Clark, J. L., & Nazzaro, S. G. 4591 (28); 4595 (26).
Clark, J. L., et al. 127 (30); 213 (5); 1178 (9); 1587 (6c); 1597 (34); 1616 (34); 1654 (5); 2322 (6c); 2410 (20); 2412 (26); 2414 (9); 2423 (15); 2424 (28); 2425 (28); 2484 (6c); 2489 (6c); 2497 (6c); 2513 (6c); 2641 (6c); 2716 (5); 2815 (30); 2816 (30); 2845 (34); 2851 (6c); 2881 (30); 3399 (9); 3644 (26); 3822 (26); 3873 (6c); 3897 (9); 3997 (6c); 4157 (6c); 4204 (30); 4305 (6c); 4610 (34).
Cobb, L. 73 (28).
Cogollo, A. 980 (3); 1279 (3).
Cogollo, A., et al. 2770 (6a); 2775 (6a); 3562 (11); 3885 (6a); 4043 (16); 4057 (11).
Contreras, E. 10891 (26).
Córdova, J. E. 1 (22).
Core, E. L. 614 (6a); 615 (16).
Cornejo, X., & Bonifaz, C. 127 (6c); 369 (6c); 1219 (6c); 2849 (6c); 3356 (6c); 4245 (26); 4322 (6c); 4823 (32); 4831 (21); 4832 (6c); 5108 (6c).
Cornejo, X., et al. 3506 (26); 4953 (7); 4969 (32).
Croat, T. B. 15957 (35); 16771 (1); 17231 (1); 19429 (9); 25304 (1); 26176 (11); 26197 (1); 26209 (11); 33292 (35); 35287 (11); 36990 (1); 38775 (19); 38876 (28); 41271 (26); 47178 (26); 49338 (28); 49790 (1); 50233 (28); 50637 (19); 50877 (21); 50878 (9); 50890 (6c); 56733 (29); 56985 (6c); 60403 (1); 66640 (35); 68111 (35); 68915 (11); 69606 (26); 69606A (8); 69984 (6a); 70864 (11); 71478 (8); 72137 (28); 72593 (9); 73002 (21); 73132 (6c); 73166 (9); 73736 (28); 74918 (35).
Croat, T. B., & Folsom, J. P. 34022 (35); 34023 (35).
Croat, T. B., & Grayum, M. H. 60430 (35); 68235 (35).
Croat, T. B., & Guanghua Zhu 76505 (35).
Croat, T. B., & Porter, D. M. 16378 (1).
Croat, T. B., & Rodríguez B., J. 61477 (9).
Cuadros, H. 1015 (6a).
Cuatrecasas, J. 11088 (9); 11409 (9); 11474 (26); 12866 (13); 12989 (13); 13005 (13); 15490 (6a); 15579 (16); 22277 (6a); 22747 (16); 23479 (26); 23978 (29).
Cuatrecasas, J., & Willard, L. 26158 (11); 26184 (11).
D'Arcy, W. G. 16018 (35); 16369B (1).
D'Arcy, W. G., et al. 13341 (1); 15989 (1).
Davalos 10 (6c).
Davidse, G., & D'Arcy, W. G. 10081 (12).
Davidson, C., & Jones, J. 9375 (35).
Davidson, M. E. 381 (35).
Davis, E. W. 508 (26).
Devia A., W. 594 (29).
Dodson, C. H. 7297 (7); 7350 (21); 7442 (10); 7445 (6c); 11595 (14).
Dodson, C. H., & Dodson, H. C. 6809 (14); 6838 (4).
Dodson, C. H., & Dodson, P. M. 11833 (31); 11834 (9); 16582 (6c).
Dodson, C. H., & Dodson, T. A. 15867 (14); 15918 (4).
Dodson, C. H., & Gentry, A. 9703 (28); 10284 (32).
Dodson, C. H., & Neill, D. 15538 (21).
Dodson, C. H., & Tan, K. 5383 (19).
Dodson, C. H., & Thien, L. B. 1054 (28); 1111 (28); 1217 (19); 1906 (26); 2075 (35).
Dodson, C. H., et al. 7117 (14); 7515 (7); 7535 (21); 7609 (4); 8446 (7); 8908 (6c); 8909 (6c, 7); 9086 (6c); 10561 (31); 14233 (28); 14248 (31); 14530 (21); 14684 (6c); 14930 (9); 15748 (28).
Dorr, L. J., & Barnett, L. C. 6090 (28); 6097 (26).
Dorr, L. J., & Valdespino, I. 6329 (9).
Dressler, R. L. 2984 (1); 4055 (1); 4200 (11); 4220 (11); 4277 (11); 4388 (1); 4924 (28); 5006 (35); 5354 (35); 5671 (11).
Dressler, R. L., & Andrews, R. 4171 (11); 4172 (11).
Dryer, V. J. 513 (35).
Dudley, T. R. 10495 (35); 11869 (35); 11931 (35).
Duke, J. A. 5331 (11); 5458 (11); 11184 (11); 11467 (11); 13600 (11); 16080 (9).
Duke, J. A., & Elias, T. 13683 (35).
Dunn, R. W. 9405011 (35); 9405036 (9).
Duque-Jaramillo, J. M. 2552 (9).
Dwyer, J. D. 2899 (18); 4387A (18); 4900 (18); 10303 (6c); 14406 (26); 14444 (26).
Dwyer, J. D., & Correa A., M. 8851 (1).
Dwyer, J. D., & Simmons, J. E. 9745 (9).
Echeverri, D. L., et al. 511 (16); 512 (11).
Ellenberg, H. 3208 (26).
Escobar, L. A. de 769 (7); 870 (6c).
Espejo, A. 2255 (26).
Espina, J., et al. 1703 (11); 2781 (11); 3726 (11).
Ewan, J. A. 16073 (28).
Fallen, M. 863 (6c).
Fegan et al. 15 (9).
Feinsinger, P., & Pounds, W. Z. 465 (35); 643 (35); 705 (35).
Fernández A., J. L., et al. 9054 (11); 9149 (11); 9588 (11); 10116 (11); 10168 (11); 10228 (6a); 10713 (35); 10716 (9).

Fernández, A. 258 (11).
Fernández, C., & Miller, J. 10 (6a).
Filskov, P., et al. 37090 (28).
Folsom, J. P. 3725 (1); 3767 (1); 4310 (35); 4367 (35); 4371 (11); 5292 (35); 10620 (11).
Folsom, J. P., & Mauseth, J. D. 8514 (11).
Folsom, J. P., et al. 4186 (11); 4822 (35).
Fonnegra, R., et al. 1235 (11); 2209 (11); 3513 (2); 3621 (11).
Forero, E., & Gentry, A. 706 (11).
Forero, E., et al. 2327 (6a); 2389 (11); 2839 (6a); 3036 (6a); 3354 (16); 5880 (11); 6232 (6a); 7306 (16).
Fosberg, F. R. 21066 (28).
Foster, R. 2786 (1); 2810 (11); 3838 (9); 9118 (35).
Franco, P., et al. 1154 (6a); 3422 (11).
Fraume, M. de et al. 150 (6a); 218 (6a).
Freiberg, M. 96006 (33); 96022 (19); 96214 (20); 96238 (17); 96263 (25).
Freire F., A. 463 (22).
Fuchs, H. P., & Zanella, L. 21915 (11).
Funk, V. A. 10507 (35).
Galdames, C., & Guerra, C. 1907 (1).
Galdames, C., et al. 3039 (1); 3146 (35).
Galeano, G., et al. 1987 (6a); 2541 (6a).
Gentry, A. 2746 (1); 4780 (11); 9541 (6c); 9745 (9); 10116 (6c).
Gentry, A., & Emmons, L. 39661 (9).
Gentry, A., & Fallen, M. 17295 (11); 17563 (16).
Gentry, A., & Forero, E. 7265 (11).
Gentry, A., & Keating, P. 59685 (20); 59693 (15); 59716 (26).
Gentry, A., & Loor C., J. 30778 (32).
Gentry, A., & Miller, J. 54943 (9).
Gentry, A., & Rentería A., E. 24542 (6a).
Gentry, A., & Shupp, G. 26391 (28).
Gentry, A., & Valencia, R. 73195 (28).
Gentry, A., et al. 12125 (19); 16783 (35); 16925 (11, 35); 25445 (9); 27463 (9); 28663 (35); 30545 (6c); 30550 (20); 30566 (26); 30569 (15); 30570 (26); 31356 (9); 34980 (20); 34988 (26); 35147 (15); 36655 (9); 37069 (11); 54264 (9); 60392 (26); 60422 (20); 60481 (28); 61790 (9); 63588 (8); 63601 (20); 65836 (9); 69962 (28); 72861 (6c); 76158 (6a).
Gilli, A. 151 (28).
Gilmartin, A. J. 361 (9); 387 (6c).
Giraldo, R. 11 (15); 36 (15); 74 (15); 151 (8).
Giraldo-Gensini, J. 293 (6a).
Giraldo-Gensini, J., & Agredo, L.O. 447 (6a).
Gómez, A., et al. 483 (11).
Gómez, J. A. 572 (11).
Gómez, L. D. 18785 (35); 19223 (35).
Gómez, L. D., & Herrera, G. 23491 (18); 23515 (18).
Gómez, L. D., et al. 20321 (18).
Gómez-Laurito, J. 10463 (18); 10599 (35).
Gómez-Laurito, J., & Vargas, G. 11960 (18).
Gómez-Laurito, J., et al. 11848 (35).
González M., E. 218 (6c).
González, M. S. 271 (6c).
Gordon, B. L. 33C (11); 65C (11).
Grant, J. R., & Rundell, J. R. 92-1921 (24).
Grayum, M., & Burton, W. 4317 (1).
Grijalva, A., et al. 604 (29).
Grijalva, E. 526 (19).
Grubb, P. J., et al. 1087 (26); 1126 (26); 1594 (35).
Gudiño, E., & Cuamacás, B. 1954 (8).
Guerra L., A. 7 (6a).
Haber, W. A., & Bello C., E. 5789 (35); 5808 (35); 5876 (35); 7052 (35).
Haber, W. A., & Zuchowski, W. 9351 (35).
Haber, W. A., et al. 4562 (35); 5331 (35); 5385 (35).
Hammel, B. 2083 (35); 3148 (11); 3213 (11); 3214 (1); 3788 (1); 5219 (11); 5802 (35); 5836 (35); 6126 (35); 7156 (1); 13637 (35).
Hammel, B., & Nevers, G. de 13587 (1).
Hammel, B., & Nepokroeff, M. 18256 (24).
Hammel, B., & Wilder, G. 16062 (28); 17238 (9).
Hammel, B., et al. 4930 (12); 4951 (1).
Hampshire, R. J., & Whitefoord, C. 178 (35); 296 (35); 510 (35).
Hansen, B., et al. 7816 (32); 7826 (32).
Harling, G., & Andersson, L. 11532 (28); 11543 (28); 11743 (9); 12334 (20); 12349 (15); 12496 (26); 14339 (6c); 16565 (9); 16603 (9); 16628 (9); 16728 (6c); 16735 (6c); 17406 (9); 17506 (9); 17579 (9); 17722 (9); 18907 (32); 19038 (21); 19332 (6c); 19407 (6c); 23067 (28); 23355 (9); 23862 (26); 23933 (26).
Harling, G., et al. 6989 (9); 7351 (9); 7606 (9); 7765 (9); 14750 (35); 14797 (6b); 15496 (6c); 19744 (9); 19985 (28).
Hatheway, W. H. 1559 (16).
Haught, O. 2913 (6c); 3109 (6c); 3205 (28); 3358 (6c); 4619 (11); 4654 (11); 5195 (26); 5214 (9); 5521 (11).
Heilborn, O. 757 (19).
Heinrichs, E. 303 (9).
Hekker, F., & Hekking, W. H. A. 10240 (6c).
Heppner, U. 3 (9).
Heredia, L. 13 (6c).
Hernández, C. 13 (6c).
Herrera, G. 3048 (18); 5949 (35); 7032 (11).
Herrera, H. 678 (11).
Herrera, H., et al. 1241 (11).
Hirtz, A. 2159 (15).
Hitchcock, A. S. 21121 (6c); 21820 (35).
Hodge, W. H. 6662 (6a); 6998 (11).
Holm-Nielsen, L. 16132 (19); 19195 (9).
Holm-Nielsen, L., & Jaramillo, J. 23331 (26).
Holm-Nielsen, L., & Jeppesen, S. 904 (9).
Holm-Nielsen, L., et al. 2899 (27); 3799 (35); 7097 (26); 21493 (9); 22010 (9); 22075 (9); 22198 (9); 22201 (9); 22300 (9); 22346 (9); 22354 (9); 22364 (9); 22443 (9); 22562 (9); 22620 (9);

22630 (9); 22638 (9); 22683 (9); 22735 (9); 25466 (6c); 25504 (9); 25551 (6c); 25611 (6c); 26919 (26); 27103 (26); 27407B (26).
Hoover, W. S. 1034 (9); 1060 (28); 1950 (26); 1972 (26); 2020 (15); 2119 (9); 2143 (20); 2240 (20); 2283 (8).
Hoover, W. S., & Wormley, S. 1424 (28); 1525 (26); 1526 (28); 1630 (28); 1775 (20); 1776 (20); 1812 (26); 1846 (9).
Hoover, W. S., et al. 2449 (9); 2468 (8); 2470 (9); 2599 (9); 3415A (9); 3597 (9); 3605 (9); 3846 (9); 4134 (9).
Hugh-Jones, D. Ll. 397 (16).
Humbles, J. 6213 (9).
Humboldt, F. H. A. von & Bonpland, A. 1878 p.p. (6a).
Hurtado, F., & Alvarado, A. 1107 (26).
Hurtado, F., & Pauchi, B. 2440 (9).
Hurtado, F., et al. 1416 (19).
Iltis, H. H., et al. E114 (4).
Jameson, W. 113 (28).
Jaramillo, J. 5152 (9); 5965 (22); 5965A (15); 5968 (31); 5972 (26); 6318 (9); 6390 (6c); 6419 (9); 7003 (10); 7031 (6c); 7630 (9).
Jaramillo, J., & Cachago, A. 4749 (26).
Jaramillo, J., & Coello, F. 2183 (9); 2647 (9); 2682 (9); 2722 (9); 3013 (9); 3275 (9); 3467 (9); 3741 (6b); 4246 (9); 4291 (9); 4643 (9).
Jaramillo, J., & Grijalva, E. 11436 (9); 11503 (9); 14528 (22).
Jaramillo, J., & Rivera, S. 139 (9); 200 (6b).
Jaramillo, J., & Winnerskjold, V. 5597 (35).
Jaramillo, J., & Zak, V. 574 (28); 577 (15); 7875 (19); 7773 (15); 7997 (31).
Jaramillo, J., et al. 5226 (9); 12349 (35); 13873 (9).
Játiva, C., & Epling, C. 522 (6c); 744 (34); 1037 (6c); 1110 (6c); 2010 (34); 2020 (9).
Johnson, H. 165 (26).
Johnston, J. R. 1849 (26).
Jørgensen, P. M., et al. 65283 (9).
Kalbreyer, W. 1509 (6a); 1511 (16); 1511A (16); 1659 (16); 1787 (16).
Kennedy, H., & Dressler, R. L. 1275 (1); 1281 (1); 3471 (1).
Killip, E. P., & Garcia, H. 33486 (11); 33818 (6a); 33835 (35).
Killip, E. P., & Hazen, T. E. 8798 (6a); 8816 (6a).
Killip, E. P., & Smith, A. C. 25759 (35).
Kirkbride, J. H., Jr. 240 (1).
Kirkbride, J. H., Jr., & Bristan, N. 1412 (11).
Kirkbride, J. H., Jr., & Chamba R., H. 4029 (26).
Kirkbride, J. H., Jr., & Duke, J. A. 432 (1); 450 (18); 508 (18); 516 (18); 573 (18); 706 (35); 791 (1); 886 (35); 1387 (11).
Klevera, J. M. 71 (11).
Klitgård, B. B. 67038 (7).
Knapp, S. 979 (1); 1421 (12); 5025 (35).
Knapp, S., & Kress, W. J. 4300 (11).
Knapp, S., & Mallet, J. 3074 (11); 3095 (11).
Knapp, S., & Schmalzel, R. 3617 (1).
Knapp, S., et al. 4147 (35); 4600 (1); 4714 (11).
Knudsen, J., & Tollsten, L. 48 (6c).
Knudsen, J., et al. 565 (6c).
Korning, J., & Thomsen, K. 47002 (6b); 47108 (6b).
Kress, W. J., & Cablk, L. D. 94-3845 (35).
Kvist, L. P. 40665 (6c); 40701 (10); 60318 (26); 60323 (35); 60327 (9); 60338 (9); 60454 (9).
Kvist, L. P., & Asanza, E. 40327 (6c); 40734 (6c).
Kvist, L. P., & Barfod, A. 49078 (31); 49079 (28).
Kvist, L. P., & Holm-Nielsen, L. 40233 (6c); 40248 (6c).
Kvist, L. P., et al. 48999 (28); 60097 (28); 60349 (35).
Laegaard, S. 51602 (9).
Lasso, S. 35 (9).
Lawesson, J. E., et al. 39424 (6b); 39587 (9); 39792 (9).
LeDoux, D. 2651 (35).
Lehmann, F. C. 1333 (26); 3228 (6a); 5157 (8); 5846 (28); 6462 (6c).
Lellinger, D. B., & White, J. J. III 1487 (35).
León, H., et al. 1269 (11); 1389 (9); 1390 (6c); 1676 (28).
Lewis, W. H., et al. 1755 (1); 13655 (9); 13823 (9).
Liesner, R., & Judziewicz, E. J. 14685 (35); 14847 (35);
Liesner, R., et al. 15545 (35).
Londoño, X., & Kvist, L. P. 128 (6b); 181 (16).
Londoño, X., & Lozano, H. 224 (26); 226 (35).
Londoño, X., et al. 494 (28).
Lozano C., G., & Díaz, J. 3105 (6a).
Lozano C., G., & Fernández, J., et al. 6875 (6a); 6887 (35).
Lozano C., G., et al. 4988 (6a).
Luer, C., & Hirtz, A. 6902 (9).
Luer, C., & Luer, J., et al. 3536 (35).
Luer, C., et al. 2673 (15); 3365 (15); 4413 (26); 4555 (26); 4714 (28); 4735 (15); 4743 (31); 7367 (28); 7381 (19); 9049 (16).
Lugo S., H. 38 (35); 275 (9); 809 (26); 981 (9); 997 (9); 1505 (9); 1564 (9); 1604 (9); 2470 (35); 2542 (9); 2658 (9); 2699 (9); 2880 (9); 3287 (9); 3500 (6b); 3548 (6b); 3560 (6b); 4096 (6b); 4195 (6b); 4627 (35); 4771 (35); 5155 (35); 5254 (6b); 5436 (9); 5871 (9); 6090 (35).
Lundell, C. L., & Contreras, E. 20953 (26).
Luteyn, J. L., & Berg, C. C. 14359 (31); 14360 (15); 14361 (26); 14366 (28).
Luteyn, J. L., & Borchsenius, F. 13340 (31).
Luteyn, J. L., & Cotton, E. 10873 (15).
Luteyn, J. L., & Lebrón-Luteyn, M. 5645 (31); 6521 (28).
Luteyn, J. L., & Sylva S., D. S. 13936 (20).
Luteyn, J. L., et al. 8444 (26); 8702 (9); 9072 (9); 10491 (6a); 12141 (11); 12545 (35); 12553 (6a).
Maas, P. J. M., & Cobb, L. 4821 (6c); 4824 (28).

Rueda, R., et al. 1022 (9).
Ruiz, H., & Pavón, J. 18/69 (6c).
Sánchez V., I., & Dillon, M. O. 8542 (35).
Sánchez, D., et al. 15 (2); 649 (11); 1021 (2); 1476 (35); 1480 (11).
Schlim, L. 543 (6a).
Schultes, R. E., & Black, G. A. 8329A (9).
Schultes, R. E., & Villarreal, M. 5220 (26); 5244 (26).
Schunke, J. M. 198 (9).
Schunke V., J. 6822 (9); 10179 (35); 11257 (35); 11864 (35); 12580 (9).
Schwerdtfeger, M. 46 (20); 69 (28).
Seemann, B. 1054 (11).
Shank, P. J., & Molina R., A. 4465 (1); 4550 (1).
Silverstone-Sopkin, F. A., et al. 2980 (6a).
Skog, L. E. 5527 (1).
Skog, L. E., & Linett, L. 5537 (1).
Skog, L. E., & Skog, J. E. 5199 (35).
Skog, L. E., et al. 4060 (35); 4180 (12); 4185 (1); 5026 (35); 5031 (35); 5103 (35); 5122 (35); 5163 (35); 5172 (35).
Skutch, A. F. 1827 (26); 3752 (35); 4648 (35); 5381 (24).
Smith, F., et al. 10899 (35).
Smith, J. D. 1798 (26).
Smith, J. F., & Galeano, M. 1510 (20).
Smith, J. F., et al. 1314 (29).
Sneidern, K. von 478 (16); 762 (16); 793 (16); 956 (9); 957 (9); 970 (9); 4521 (9); 5075A (6a); 5464 (16).
Snow, D. W. 18 (26); 22 (8).
Sobel, G., & Strudwick, J. 2446 (35).
Sodiro, L. 119/50 (26); 119/53 (28); 119/55 (9).
Solomon, J. 14838A (35).
Spruce, R. 5069 (6b).
Ståhl et al. 2119 (32).
Standley, P. C. 68121 (26); 68145 (26); 68578 (26); 71311 (26); 85056 (26); 85456 (26); 85544 (26); 86410 (26); 90664 (26).
Standley, P. C., & Valerio, J. 46867 (35); 49757 (35); 49788 (35); 49806 (35); 49813 (35); 50054 (35); 50289 (35); 50989 (35); 51798 (35); 52016 (35).
Stein, B. A. 2622 (35); 2948 (26).
Stein, B. A., & Suárez, A. 3044 (9).
Stein, B. A., & Tucker, D. 3110 (9).
Stevens, W. D. 13792 (35).
Steyermark, J. A. 34290 (26); 34336 (26); 35201 (26); 36353 (26); 36796 (26); 36995 (26); 52577 (28); 54215 (7); 54225 (6c).
Stiles, F. G. 563 (28); 570 (8); 571 (26); 574 (9); 575 (15); 582 (20).
Stolze, R. G. 1573 (35).
Sullivan, G. A. 146 (1).
Sytsma, K. J. 1639 (12); 1682 (1).
Sytsma, K., & Andersson, L. 4647 (35).
Sytsma, K., et al. 4273 (1); 4336 (1).
Tate, G. H. H. 476 (34); 513 (6c).
Taylor, J., & Taylor, C. 11506 (35).
Tessmann, G. 3720 (9); 4991 (9).
Thomas, W. W., & Rios, M. 6665 (9).
Thomsen, K. 374 (24).
Timaná, M. 748 (35).
Tipaz, G., et al. 2026 (9).
Tirado, M., et al. 233 (6c); 705 (9); 1278 (26); 1283 (28).
Tonduz, A. 9192 (18); 9554 (1); 13038 (18).
Troya, N. 24 (6c).
Tunqui, S. 681 (9); 715 (9).
Türckheim, H. von 196 (26); II1740 (26); II1743 (26); II2281 (26).
Ule, E. 6671 (9).
Utley, J., & Utley, K. 1112 (24).
Valerio R., J. 1246 (35).
Vargas C., C. 6794 (35); 16844 (35).
Vargas, H., & Grefa, P. 880 (9).
Vargas, H., & Sandoval, D. 459 (26).
Vásquez, R., & Jaramillo, N. 6332 (9); 6338 (9); 9733 (9); 9900 (9); 10170 (9).
Ventura, E., & López, E. 3125 (26).
Vieira, J., & Varas C., 5 (9).
Villavicencio F., H. 10 (6c).
Vogel, S. 93 (29).
Warner, R. H., & White, J. W. 129 (11).
Watson, J. B. 160 (6c).
Webster, G. L., & Delprete, P. 27236 (28).
Webster, G. L., et al. 27125 (28); 27134 (19); 27274 (19); 27794 (6c); 27896 (28); 30038 (19); 30265 (28).
Wedel, H. von 300 (35); 1420 (1); 1475 (1); 2272 (1); 2346 (1); 2347 (1); 2351 (1); 2455 (1).
Wendland, H. 568 (35); 922 (35).
Werff, H. van der & Gudiño, E. 11209 (9).
Werff, H. van der et al. 12053 (6c); 12206 (19); 12285 (28); 12387 (7); 12394 (4).
Werling, L., & Leth-Nissen, S. 359 (15); 393 (28); 547 (6c).
White, J., & Warner, R. H. 15 (11).
Wiehler, H. 7111 (6c); 7157 (9); 71118 (9); 71161 (11); 71162 (1); 71270 (1); 72190 (9); 72312 (1); 72316 (1); 78167 (11).
Wiehler, H., & Dressler, R. L. 71165 (12).
Wiggins, I. L. 10946 (6c).
Williams, L. O. 19515 (18).
Williams, L. O., et al. 25690 (26); 26235 (26); 26270 (26); 26950 (26); 27179 (26); 28588 (35).
Wilson, R. 65272 (35).
Wittmann, U. 71 (9).
Wittmann, U., & Arroyo, S. 10 (6c).
Woodson, R. E. Jr., & Schery, R. W. 669 (35).
Wurdack, J. J. 2277 (6b).
Yánez, A. P., et al. 1600 (9).
Zak, V., & Espinoza, S. 5284 (6b).
Zak, V., & Jaramillo, J. 574 (28); 577 (15).
Zak, V., et al. 5369 (6c); 5538 (6c); 5551 (6c).
Zarucchi, J. L., et al. 5513 (11); 6051 (6a).

INDEX TO SCIENTIFIC NAMES

Accepted names are in roman type; the main entry for each is in **boldface**. Synonyms are in *italics*.

SYSTEMATIC BOTANY MONOGRAPHS

Vol. 26. Revision of Palicourea (Rubiaceae) in Mexico and Central America, Charlotte M. Taylor, 102 pp, 1989. ISBN 0-912861-26-6. US orders: $12.50; non-US orders: $13.50.

Vol. 27. Systematics of Chusquea section Swallenochloa, section Verticillatae, section Serpentes, and section Longifoliae (Poaceae-Bambusoideae), Lynn G. Clark, 127 pp, 1989. ISBN 0-912861-27-4. US orders: $15.00; non-US orders: $16.00.

Vol. 28. Systematics of Oropogon (Alectoriaceae) in the New World, Theodore L. Esslinger, 111 pp, 1989. ISBN 0-912861-28-2. US orders: $13.50; non-US orders: $14.50.

Vol. 29. Revision of Hieracium (Asteraceae) in Mexico and Central America, John H. Beaman, 77 pp, 1990. ISBN 0-912861-29-0. US orders: $9.00; non-US orders: $10.00.

Vol. 30. Systematics of Simsia (Compositae-Heliantheae), David M. Spooner, 90 pp, 1990. ISBN 0-9128612-30-4. US orders: $11.00; non-US orders: $12.00.

Vol. 31. Systematics of the annual species of Muhlenbergia (Poaceae-Eragrostideae), Paul M. Peterson and Carol R. Annable, 109 pp, 1991. ISBN 0-912861-31-2. US orders: $13.50; non-US orders: $14.50.

Vol. 32. Systematics of Clermontia (Campanulaceae-Lobelioideae), Thomas G. Lammers, 97 pp, 1991. ISBN 0-912861-32-0. US orders: $13.00; non-US orders: $14.00.

Vol. 33. Taxonomy of Complaya, Elaphandra, Iogeton, Jefea, Wamalchitamia, Wedelia, Zexmenia, and Zyzyxia (Compositae-Heliantheae-Ecliptinae), John L. Strother, 111 pp, 1991. ISBN 0-912861-33-9. US orders: $14.50; non-US orders: $15.50.

Vol. 34. Systematics of Epilobium (Onagraceae) in China, Chia-jui Chen, Peter C. Hoch, and Peter H. Raven, 209 pp, 1992. ISBN 0-912861-34-7. US orders: $25.00; non-US-orders: $26.00.

Vol. 35. Revision of Crataegus section Crataegus and nothosection Crataeguineae (Rosaceae-Maloideae) in the Old World, Knud Ib Christensen, 199 pp, 1992. ISBN 0-912861-35-5. US orders: $25.00; non-US-orders: $26.00.

Vol. 36. Systematics of Pappobolus (Asteraceae-Heliantheae), Jose L. Panero, 195 pp, 1992. ISBN 0-912861-36-3. US orders: $25.00; non-US-orders: $26.00. [Jesse M. Greenman Award, 1993]

Vol. 37. Biogeography and Systematics of Poitea (Leguminosae), Matt Lavin, 87 pp, 1993. ISBN 0-912861-37-1. US orders: $11.00; non-US-orders: $12.00. [Henry Allan Gleason Award, 1993]

Vol. 38. Monograph of Desmanthus (Leguminosae-Mimosoideae), Melissa Luckow, 166 pp, 1993. ISBN 0-912861-38-X. US orders: $20.00; non-US-orders: $21.00. [Jesse M. Greenman Award, 1994]

Vol. 39. Monograph of Mecranium (Melastomataceae-Miconieae), James D. Skean, Jr., 116 pp, 1993. ISBN 0-912861-39-8. US orders: $15.00; non-US-orders: $16.00.

Vol. 40. Revision of Cymbopetalum and Porcelia (Annonaceae), Nancy A. Murray, 121 pp, 1993. ISBN 0-912861-40-1. US orders: $16.00; non-US-orders: $17.00.

Vol. 41. Revision of Passiflora subgenus Decaloba section Pseudodysosmia (Passifloraceae), John M. MacDougal, 146 pp, 1994. ISBN 0-912861-41-X. US orders: $19.00; non-US-orders: $20.00.

Vol. 42. Revision of Timonius subgenus Abbottia (Rubiaceae-Guettardeae), Steven P. Darwin, 86 pp, 1994. ISBN 0-912861-42-8. US orders: $11.00; non-US-orders: $12.00.

Vol. 43. Revision of the Mexican species of Cyperus (Cyperaceae), Gordon C. Tucker, 213 pp, 1994. ISBN 0-912861-43-6. US orders: $28.00; non-US-orders: $29.00.

Vol. 44. Systematics of Columnea section Pentadenia and section Stygnanthe (Gesneriaceae), James F. Smith, 89 pp, 1994. ISBN 0-912861-44-4. US orders: $11.00; non-US-orders: $12.00.

Vol. 45. Phylogenetic systematics and biogeography of the tribe Robinieae (Leguminosae), Matt Lavin and Mario Sousa S., 165 pp, 1995. ISBN 0-912861-45-2. US orders: $21.00; non-US-orders: $22.00.

Vol. 46. Systematics of Cryosophila (Palmae), Randall J. Evans, 70 pp, 1995. ISBN 0-912861-46-0. US orders: $9.00; non-US-orders: $10.00.

Vol. 47. Revision of Philodendron subgenus Pteromischum (Araceae) for Pacific and Caribbean tropical America, Michael H. Grayum, 233 pp, 1996. ISBN 0-912861-47-9. US orders: $30.00; non-US-orders: $32.00.

Vol. 48. Revision of Houstonia (Rubiaceae-Hedyotideae), Edward E. Terrell, 118 pp, 1996. ISBN 0-912861-48-7. US orders: $15.00; non-US-orders: $16.00.

Vol. 49. Revision of Geranium subgenus Erodioidea (Geraniaceae), Carlos Aedo, 104 pp, 1996. ISBN 0-912861-49-5. US orders: $14.00; non-US-orders: $15.00.

Vol. 50. Systematics of Oenothera section Oenothera subsection Oenothera (Onagraceae), Werner Dietrich, Warren L. Wagner, and Peter H. Raven, 234 pp, hardbound, color frontispiece, 1997. ISBN 0-912861-50-9. US orders: $30.00; non-US-orders: $32.00.

Vol. 51. Monograph of Stigmaphyllon (Malpighiaceae), Christiane Anderson, 313 pp, hardbound, color frontispiece, 1997. ISBN 0-912861-51-7. US orders: $40.00; non-US-orders: $42.00.

THE AMERICAN SOCIETY OF PLANT TAXONOMISTS

http://www.sysbot.org

The American Society of Plant Taxonomists publishes the journal *Systematic Botany* four times a year and the series *Systematic Botany Monographs* at irregular intervals. For information concerning the journal, institutional subscriptions, and membership in the Society contact the Business Manager, Linda J. Brown, Department of Botany, University of Wyoming, Laramie, WY 82071-3165 USA; telephone: 307-766-2556; fax: 307-766-2851; e-mail: aspt@uwyo.edu.

SYSTEMATIC BOTANY MONOGRAPHS

The ASPT publishes taxonomic monographs and revisions, longer than 50 printed pages, in *Systematic Botany Monographs*. Instructions for authors are available from the editor, Christiane Anderson (telephone: 734-647-2812; fax: 734-647-5719; e-mail: chra@umich.edu). Standing orders and orders for individual volumes should be sent to:

Systematic Botany Monographs
University of Michigan Herbarium
North University Building
Ann Arbor, Michigan 48109-1057 U.S.A.

Terms of Sale:

Standing orders receive a 10% discount and are billed when each volume is mailed. Orders for individual volumes must be paid in advance of shipment, and no discount is allowed; an invoice will be sent on request. All payments must be in U.S. currency. Make checks payable to *The American Society of Plant Taxonomists*. Orders are mailed postpaid. Mastercard and Visa orders accepted; include account number and expiration date. Not available as exchange.

PREVIOUSLY PUBLISHED VOLUMES (CONTINUED FROM COVER 3):

Vol. 52. Infrageneric classification of Salix (Salicaceae) in the New World, George W. Argus, 121 pp, 1997. ISBN 0-912861-52-5. US orders: $15.00; non-US-orders: $16.00.

Vol. 53. Revision of Cuphea Section Diploptychia (Lythraceae), Shirley A. Graham, 96 pp, 1998. ISBN 0-912861-53-3. US orders: $13.00; non-US-orders: $14.00.

Vol. 54. Monograph of Kadsura (Schisandraceae), Richard M. K. Saunders, 106 pp, color frontispiece + 1 color plate, 1998. ISBN 0-912861-54-1. US orders: $14.00; non-US-orders: $15.00

Vol. 55. Monograph of Leucaena (Leguminosae-Mimosoideae), Colin Hughes, 244 pp, hardbound, 1998. ISBN 0-912861-55-X. US orders: $32.00; non-US-orders: $34.00 [Prix Augustin-Pyramus de Candolle, 1998]

Vol. 56. Monograph of Pictetia (Leguminosae-Papilionoideae) and review of the Aeschynomeneae, Angela Beyra M. and Matt Lavin, 93 pp, 1999. ISBN 0-912861-56-8. US orders: $13.00; non-US-orders: $14.00.

Vol. 57. Systematics of Cobaea (Polemoniaceae), L. Alan Prarther, 81 pp, 1999. ISBN 0-912861-57-6. US orders: $11.00; non-US-orders: $12.00.

Vol. 58. Monograph of Schisandra (Schisandraceae), Richard M. K. Saunders, 146 pp, color frontispiece + 5 color plates, 2000. ISBN 0-912861-58-4. US orders: $20.00; non-US orders: $21.00.

Vol. 59. Revision of Gasteranthus (Gesneriaceae), Laurence E. Skog and Lars P. Kvist, 118 pp, 2000. ISBN 0-912861-59-2. US orders: $15.00; non-US orders: $16.00.